GENETICS

PRACTICE PROBLEMS AND SOLUTIONS

JOSEPH CHINNICI
Virginia Commonwealth University

DAVID MATTHES
San Jose State University

An imprint of Addison Wesley Longman, Inc.

Menlo Park, California • Reading, Massachusetts • New York
Harlow, England • Don Mills, Ontario • Sydney
Mexico City • Madrid • Amsterdam

Project Editor: Evelyn Dahlgren
Senior Production Editor: Larry Olsen
Production Supervisor: Brian Jones
Copyeditor: Sylvia Stein Wright
Composition: Jupiter Productions
Artists: Mark Konrad, Menocino Graphics
Cover Design: Karl Miyajima

Figure Credits

14.2 Adapted from W.B. Wood and H.R. Revel, "The Genome of Bacteriophage T4, *Bacteriological Review* 40:847-868. Reprinted by permission.
15.2 Adapted from an illustration by Irving Geiss.
19.4 From Gunter Kahl, *Dictionary of Gene Technology* (VCH, 1995), p. 376.
21.12 Reprinted from *The Genetics and Biology of Drosophila*, Vol. 1a, Ashburner and Novitski, eds., pp. 31-66. Copyright © 1992, and used by permission of the publisher, Academic Press.

ISBN 0-8053-4525-6

1 2 3 4 5 6 7 8 9 10–CRS–03 02 01 00 99 98

The Benjamin/Cummings Publishing Company, Inc.
2725 Sand Hill Road
Menlo Park, California 94025

Contents

Preface

Most educators today agree that students who are actively involved in the learning process will master a subject more thoroughly and meaningfully and will tend to retain the information for longer periods of time. Our goal in preparing this compilation of genetics and molecular biology practice problems is to stimulate active student learning of this vitally important and intensely interesting area of modern experimental biology.

In each chapter we have included a variety of problems meant to stimulate critical analysis and interpretation of data. Each chapter has a section devoted to more challenging applications of genetics principles and a set of questions amenable to analysis by students working in cooperative groups. The solutions to all the problems are presented at the end of each chapter. Explanations are included to provide a positive and encouraging starting point for students who may not have encountered these particular applications of genetics principles.

We hope that students of genetics use this book as an opportunity for review and practice. Only after working on a problem for a while, consulting your textbook, and writing down an answer should you consult the solution. As with most things in life, the learning occurs when your mind is engaged in the struggle to find a solution. We hope you enjoy solving the problems in this book and thereby come to appreciate the power and beauty of genetic analysis.

Acknowledgments

I would like to thank those geneticists who have been instrumental in teaching the value of clearly focused and crisply written genetics problems as both teaching and learning tools. For me those mentors include Professors Roland Holroyd, J. James Murray, Jr., Theordore R.F. Wright, J. Ives Townsend, and Walter E. Nance.

Joseph Chinnici

I would like to thank Dr. Bruce Chase, Leonard Lao, Jessica Ryan, Payam Shahi, and Victoria Wu for their helpful comments on early drafts of this manuscript. The careful reading and constructive suggestions by Nina Horne, Anita Bennett, and Sylvia Stein Wright are also greatly appreciated. Last and first, many thanks to my wife Devavani, whose support and encouragment helped the writing process be swift and joyous.

David Matthes

About the Authors

Dr. Joseph Chinnici studied biology and genetics at La Salle College and the University of Virginia. He is currently teaching at Virginia Commonwealth University. Dr. Chinnici teaches a course called Science of Heredity for liberal arts students as well as genetics courses for biology majors. He lives with his wife and three miniature schnauzers in Richmond, where he enjoys photography, visiting lighthouses, and attending Phillies games.

Dr. David Matthes received his Ph.D. from the University of California at Berkeley in 1995. He currently teaches courses in general and molecular genetics at San Jose State University. His research program is focused on finding the genetic similarities between axon guidance in insects and white blood cell migration in humans.

Segregation, Independent Assortment, and Probability

Problems

1. In the pea, tall plant height (T) is dominant over short (t). Pure-breeding tall and short plants are crossed.

    a. What will be the genotype and phenotype of the F_1?

    b. If the F_1 is selfed and 400 F_2 plants are raised, how many would be expected in each phenotypic class?

    c. How many of the F_2 would be expected to be pure breeding when selfed?

2. Consider a tall pea plant:

    a. If nothing is known of the breeding history of the plant, how should its genotype be written?

    b. If the plant were to be testcrossed, what would be the phenotype of the other parent?

    c. If all of the progeny from the testcross were tall, what would be the genotype of the tall plant?

    d. If the tall plant were heterozygous, what would be the phenotypic ratio of the testcross progeny?

3. The inheritance of black versus gray coat color was investigated in cats from an Australian population.

    a. Several matings between different pairs of black cats consistently produced all black progeny. Can we conclude from these results that gray is recessive?

    b. Several different matings between pairs of black and gray cats yielded both F_1 black and gray progeny. Can the mode of inheritance be determined from these data?

    c. Matings between the F_1 black cats (who are offspring of matings between black and gray parents) occasionally produced a gray kitten. Can the mode of inheritance be determined from these data?

4. The average pea pod contains about 7 peas. If heterozygous round pea plants are self-fertilized, what proportion of the pods will have:

    a. all round peas?

    b. 5 round and 2 wrinkled peas?

5. Two heterozygous tasters for PTC have a taster daughter. If the daughter marries a nontaster, what is the probability that they will have a nontaster child?

6. Polydactyly (extra digits) is a dominant trait caused by gene P, as opposed to the normal allele p. Cystic fibrosis, c, is a recessive disease, as opposed to the normal condition, C. A polydactylous woman, otherwise normal in phenotype, marries a healthy normal man. Their four children have the following phenotypes:

> Child 1 is normal in all respects.
> Child 2 is polydactylous, otherwise normal.
> Child 3 has cystic fibrosis, otherwise normal.
> Child 4 has cystic fibrosis and is polydactylous.

    a. What is the genotype of the mother?

    b. What is the genotype of child 3?

    c. What is the genotype of child 4?

    d. What is the chance that child 1 is heterozygous for cystic fibrosis?

7. The following matings were performed with mice strains differing in hair color (black or brown) and hair texture (frizzy or straight):

MATING 1		MATING 2	
female with	male with	female with	male with
straight	× straight	straight	× frizzy
black hair	black hair	black hair	brown hair
↓		↓	
straight black		straight black	
frizzy brown		frizzy brown	
straight brown		straight brown	
frizzy black		frizzy black	

    a. Which hair color and which hair texture is dominant?

    b. What are the genotypes of the parents in mating 1 and mating 2?

    c. If each set of parents produced a total of 80 offspring over their lifetimes, how many of each type would be expected? Use the forked-line method to determine your answer.

8. As Mendel discovered, tall plant height is dominant to short. In the following experiments, parents with known phenotypes but unknown genotypes produced the following progeny:

		PROGENY	
	PARENTS	TALL	SHORT
a.	tall × short	82	78
b.	tall × tall	118	39
c.	short × short	0	50
d.	tall × short	74	0
e.	tall × tall	90	0

Using the letter T for tall and t for short, give the most probable genotype of each parent.

9. In crosses b, d, and e of Problem 8, indicate how many of the tall progeny produced by each cross would be expected to produce short progeny when self-fertilized.

10. A black stallion of unknown ancestry was captured and mated to a number of sorrel (red) mares, each with a purebred pedigree. These matings produced 29 sorrel offspring and 33 black offspring.

    a. Which color is most likely caused by a recessive homozygote?

    b. According to your hypothesis, what numbers of each genotype and each phenotype would you expect?

11. In guinea pigs, dark coat color (C) is dominant over albino (c), and short hair (S) is dominant over long hair (s). The genes for these two traits show independent assortment. Write the most probable genotypes for the parents of each of the following crosses:

		PHENOTYPES OF OFFSPRING			
		SHORT DARK	LONG DARK	SHORT ALBINO	LONG ALBINO
a.	dark short × dark short	89	31	29	11
b.	dark short × dark long	18	19	0	0
c.	dark short × albino short	20	0	21	0
d.	albino short × albino short	0	0	28	9
e.	dark long × dark long	0	32	0	10
f.	dark short × dark short	46	16	0	0
g.	dark short × dark long	29	31	9	11

12. In humans, two traits, widow's peak and free-hanging earlobes, depend on separate dominant genes located on different chromosomes. A man with a widow's peak and attached earlobes (whose father has free-hanging earlobes) married a woman without a widow's peak but with free-hanging earlobes (whose father had attached earlobes). What is the probability that their first child will:

    a. not have either a widow's peak or attached earlobes?

    b. have both a widow's peak and free-hanging earlobes?

    c. have a widow's peak and attached earlobes?

    d. not have either a widow's peak or free-hanging earlobes?

13. Assume that capital letter alleles are dominant to lowercase letter alleles. From the following cross:

PARENT 1		PARENT 2
AabbCcDd	×	*AaBbCcDd*

    a. What fraction of the progeny will be *aaB–C–dd*?
    b. What fraction of the progeny will be like parent 2 in phenotype?
    c. What fraction of the progeny will be unlike either parent in phenotype?
    d. What fraction of the progeny will be *AABBCcDd*?
    e. What fraction of the progeny will be *AaBbccdd*?
    f. What fraction of the progeny will be unlike either parent in genotype?

14. In garden peas, tall plants (*T*) are dominant over short (*t*), round seeds (*R*) are dominant over wrinkled (*r*), yellow seeds (*Y*) are dominant over green (*y*), and purple flowers (*A*) are dominant over white (*a*). Consider the following pea plants, and answer the questions below:

    plant 1 = *RryyAaTt*, plant 2 = *RrYYAatt*

    In a cross of plant 1 × plant 2, what proportion of the progeny will:
    a. have the genotype *rrYyaatt*?
    b. have the phenotype wrinkled, yellow, white, short?
    c. be dominant for all four characteristics?
    d. be pure breeding (homozygous) for seed shape?
    e. be pure breeding for round *and* purple?
    f. be pure breeding for *all* four characteristics?
    g. What proportion of the round yellow white tall progeny will have the genotype *RrYyaaTt*?

15. In sesame, both the number of seed pods per leaf axil (3-pod or 1-pod) and the shape of the leaf (wrinkled or smooth) are determined by pairs of alleles at two different genetic loci, and they show independent assortment. The result of six crosses gave the results below. Fill in the full genotypes for all the parents and the generalized genotypes for the four categories of offspring:

	NUMBERS OF PROGENY			
	1-POD SMOOTH	1-POD WRINKLED	3-POD SMOOTH	3-POD WRINKLED
PARENTS	_____	_____	_____	_____
1-pod smooth × 3-pod wrinkled	211	0	205	0
1-pod wrinkled × 3-pod smooth	78	90	84	88
1-pod smooth × 1-pod smooth	447	158	146	52
1-pod smooth × 3-pod smooth	318	98	323	104
1-pod smooth × 1-pod wrinkled	110	113	33	38
1-pod smooth × 3-pod smooth	362	118	0	0

Challenge Problems

Background for Problems 16 and 17: In the pea, round seeds (R), yellow seeds (Y), purple flowers (P), and tall height (T) are dominant traits, and wrinkled seeds (r), green seeds (y), white flowers (p), and short height (t) are recessive. Consider the following cross: plant 1: *RryyPpTt* × plant 2: *RrYYPptt*.

16. If testcrosses were used to confirm the genotypes of the parental plants,
    a. what would be the phenotype of the tester plant?
    b. what phenotypes would be seen in a testcross of plant 1?
    c. what phenotypes would be seen in a testcross of plant 2?

17. In a cross of plant 1 × plant 2, what proportion of the progeny will:
    a. have the genotype *rrYypptt*?
    b. have the phenotype wrinkled, yellow, white, short?
    c. show the dominant trait for all four characteristics?
    d. be pure breeding for the characteristic of seed shape?
    e. be pure breeding for round seeds and purple flowers?
    f. be pure breeding for all four characteristics?
    g. What proportion of the round yellow white tall progeny will have the genotype *RrYyppTt*?

18. In humans, the ability to taste PTC is inherited as a dominant gene (T). In a marriage between 2 heterozygous tasters (Tt):
    a. what is the probability of 3 taster children?
    b. what is the probability of 3 taster girls?
    c. if they have 5 children, what is the probability that the first 3 will be tasters and the last 2, nontasters?
    d. how many different ways could they have 3 taster and 2 nontaster children in any order?
    e. what is the probability that they will have 3 taster and 2 nontaster children in any order?

Team Problems

19. In the spaces below, fill in the proportions of progeny of each phenotype if the listed hybrids are selfed.

HYBRID SELF-FERTILIZED	OFFSPRING AND PHENOTYPIC RATIOS							
	A–B–D–	*A–B–dd*	*A–bbD–*	*A–bbdd*	*aaB–D–*	*aaB–dd*	*aabbD–*	*aabbdd*
AABBDd	___	___	___	___	___	___	___	___
AaBBDd	___	___	___	___	___	___	___	___
AaBBDD	___	___	___	___	___	___	___	___
AaBbDD	___	___	___	___	___	___	___	___
AaBbDd	___	___	___	___	___	___	___	___

20. In the spaces below, give the *exact* genotype of the male parent in each cross, based on the types of offspring produced in each case.

GENOTYPE OF FEMALE PARENT	GENOTYPE OF MALE PARENT	ABD	ABd	AbD	Abd	aBD	aBd	abD	abd
AaBbDd	_____	9	3	3	1				
aaBbDd	_____	1	1	1	1	1	1	1	1
aaBbDd	_____	3	3	1	1	3	3	1	1
Aabbdd	_____	3	3	3	3	1	1	1	1
Aabbdd	_____	1		1		1		1	

21. Fill in the blanks in the table to indicate the mode of inheritance of gray and brown fur color and the presence or absence of white spotting in the fur. There are 25 blanks to fill in.

CROSSES PERFORMED				PROGENY				
Male Parents		Female Parents		Phenotypes and Numbers Observed (NOT RATIOS)				
Phenotype	Genotype	Phenotype	Genotype	Gray with Spots	Gray without Spots	Brown with Spots	Brown without Spots	TOTAL Number
_____	AABB	_____	AABB	180	0	0	0	180
Gray without spots	_____	Brown with spots	_____	52	47	46	54	199
_____	AaBB	_____	aaBB	___	___	___	___	202
_____	AABb	_____	aaBb	___	___	___	___	196
Brown with spots	_____	Brown with spots	_____	0	0	151	48	199
Brown _____	_____	_____ with spots	_____	70	24	80	26	200
Gray with spots	_____	_____	_____	115	40	36	13	204

Solutions

1. a. F_1 genotype = Tt, F_1 phenotype = tall

   Parents:　　tall TT × short tt
   $$\downarrow$$
   F_1:　　　　　Tt tall

   b. 300 tall and 100 short. Selfing the F_1 ($Tt \times Tt$) would produce a genotypic ratio 1/4 TT: 2/4 Tt: 1/4 tt. Because tall (T) is dominant, the phenotypic ratio would be 3/4 tall and 1/4 short. For 400 progeny: 3/4 × 400 = 300 tall and 1/4 × 400 = 100 short.

   c. 200. Only the homozygotes will be pure breeding. That includes the 100 short plants because they must be homozygous (1/4 cc × 400 = 100) plus the homozygous dominant tall plants (1/4 CC × 400 = 100).

2. a. $T-$, because T is dominant, the plant may be TT or Tt. The generalized genotype $T-$ is used to symbolize this situation.

   b. short, because it must have the genotype tt

   c. TT, because all the gametes produced were T

   d. 1/2 tall: 1/2 short, because half the gametes would be T and half would be t

3. a. No. Gray could be dominant (allele G) and the matings between black cats could be $gg \times gg$. However, black could be dominant (allele B), and if the black alleles occur in a high frequency in the population, most matings between black cats would involve at least one homozygous black (BB) and produce all black progeny.

   b. No. Gray may be dominant: $gg \times Gg \rightarrow Gg$ (gray) and gg (black). Or black may be dominant: $Bb \times bb \rightarrow bb$ (gray) and Bb (black).

   c. Black is dominant. The F_1s of the cross black × gray must be heterozygous Bb. Crosses between these F_1s ($Bb \times Bb$) have a 1/4 probability of producing a gray kitten (bb). The black allele occurs in such a high frequency in the Australian population that one or both of two cats picked at random are likely to be BB. Black heterozygotes were assured by using the black progeny of a cross of black × gray.

4. a. 13.5%. $(3/4)^7 = \dfrac{2{,}187}{16{,}384}$

   b. 31.2% from the binomial:
   $$n = 7$$
   $$PA = P\ R- = 3/4 : r = 5$$
   $$PB = P\ rr = 1/4 : s = 2$$

   $$\dfrac{7!}{5! \times 2!}\ [\ (3/4)^5\ (1/4)^2\ ]$$

5. 1/3. The taster daughter may be either *TT* or *Tt,* and she can have a nontaster child only if she has the recessive gene. We already know that she is a taster, so the chance that she is heterozygous must be determined by conditional probability: the probability of a heterozygote from her parents is 2/4, the probability of a taster is 3/4, and the probability that the daughter is *Tt* is 2/4 × 4/3 = 2/3. If she is heterozygous, the probability of producing a nontaster child (*tt*) by mating with a nontaster (*Tt* × *tt*) is 1/2. Therefore, the probability is determined by multiplying the probabilities of two independent events, the probability that the daughter carries the recessive gene (2/3) times the probability that the mating will produce a nontaster child (1/2) = 2/6 or 1/3.

6. a. *PpCc*

   b. *ppcc*

   c. *Ppcc*

   d. 2/3

   From their phenotypes, the genotypes of the mother and father can be written *P–C–* and *ppC–*, respectively. Because children 3 and 4 suffer from cystic fibrosis (*cc*), both parents must be heterozygous (*Cc*) for this characteristic. Children 1 and 3 are not polydactylous (*pp*). Thus, the mother must be heterozygous (*Pp*) for this characteristic. Child 3 must be homozygous recessive (*ppcc*) to have normal digits and cystic fibrosis. Child 4 must be heterozygous for polydactyly (*Pp*) because the father is homozygous recessive (*pp*) and suffers from cystic fibrosis (*cc*). Because child 1 does not suffer from cystic fibrosis (*C–*) and both parents were heterozygous (*Cc*), the chance that child 1 is heterozygous is 2/3 by conditional probability.

7. a. From mating 1, we see that frizzy hair is recessive because both parents have straight hair but some of the offspring have frizzy hair. Likewise, brown hair is recessive because both parents have black hair but some of the offspring are brown. Mating 2 does not provide information regarding dominance relationships because the parents differ in phenotype for both traits.

   b. Both parents in mating 1 are doubly heterozygous (*AaBb*) because they have dominant phenotypes for both traits but produce some recessive offspring for each trait. In mating 2, the frizzy brown male parent is recessive in phenotype for both traits and must, therefore, be homozygous (*aabb*). The female parent shows the phenotypes of both dominant traits, but must be doubly heterozygous in genotype (*AaBb*) because recessive offspring for each trait are produced.

   c. For mating 1:

   straight black female × straight black male

EXPECTED PHENOTYPIC RATIOS AND NUMBERS

Aa × *Aa*		*Bb* × *Bb*	RATIOS	NUMBERS
↓		↓		
	↗ 3/4 black	= 9/16 straight black	× 80 = 45
3/4 straight				
↗	↘ 1/4 brown	= 3/16 straight brown	× 80 = 15
↘	↗ 3/4 black	= 3/16 frizzy black	× 80 = 15
1/4 frizzy				
	↘ 1/4 brown	= 1/16 frizzy brown	× 80 = 5
			total	80

For mating 2:

straight black female × frizzy brown male

EXPECTED PHENOTYPIC RATIOS AND NUMBERS

Aa × *Aa*	*Bb* × *Bb*	RATIOS	NUMBERS
↓	↓		
	↗ 1/2 black	= 1/4 straight black	× 80 = 20
1/2 straight			
↗	↘ 1/2 brown	= 1/4 straight brown	× 80 = 20
↘	↗ 1/2 black	= 1/4 frizzy black	× 80 = 20
1/2 frizzy			
	↘ 1/2 brown	= 1/4 frizzy brown	× 80 = <u>20</u>
			total 80

8. If parents produce recessive offspring, each must have at least one recessive gene. If two dominant parents have all dominant offspring, one of them might be heterozygous.
   a. *Tt* × *tt*
   b. *Tt* × *Tt*
   c. *tt* × *tt*
   d. *TT* × *tt*
   e. *TT* × (*TT* or *Tt*)

9. b. 2/3, because the tall progeny are produced in a ratio of 1/4 *TT* and 2/4 *Tt* (a 1:2 ratio)
   d. all, because all the progeny will be *Tt*
   e. 1/2 (if *GG* × *Gg*) or none (if *GG* × *GG*)

10. a. sorrel
    b. 1/2 black and 1/2 sorrel

11. Consider each trait separately. Remember, if the parents produce recessive offspring for a particular trait, each dominant parent must be heterozygous.
    a. *CcSs* × *CcSs*
    b. *CCSs* × *CCss* (one could be *Cc*)
    c. *CcSS* × *ccSS* (one could be *Ss*)
    d. *ccSs* × *ccSs*
    e. *Ccss* × *Ccss*
    f. *CCSs* × *CCSs* (one could be *Cc*)
    g. *CcSs* × *Ccss*

12. a. 1/4
    b. 1/4
    c. 1/4
    d. 1/4

13. a. **3/128, because**

$Aa \times Aa$	$bb \times Bb$	$Cc \times Cc$	$Dd \times Dd$
↓	↓	↓	↓

$1/4\ aa$　×　$1/2\ B{-}$　×　$3/4\ C{-}$　×　$1/4\ dd$　=　$3/128$

b. **27/128**

c. **74/128, because**
$1 - [(A{-}bbC{-}D{-}) - (A{-}B{-}C{-}D{-})] = 1 - [(27/128) - (27/128)]$
$= 1 - (54/128) = 74/128$

d. **0**

e. **1/64**

f. **14/16**

14. a. **1/32**

b. **1/32**

c. **9/32**

d. **1/2**

e. **1/16**

f. **none**

g. **2/3**

15. Consider each trait separately. Remember, if the parents produce recessive offspring for a particular trait, each dominant parent must be heterozygous.

	NUMBERS OF PROGENY			
	1-POD	**1-POD**	**3-POD**	**3-POD**
	SMOOTH	**WRINKLED**	**SMOOTH**	**WRINKLED**
PARENTS	**A–B–**	**A–bb**	**aaB–**	**aabb**
1-pod smooth × 3-pod wrinkled *AaBB*　　　　*aabb*	211	0	205	0
1-pod wrinkled × 3-pod smooth *Aabb*　　　　*aaBb*	78	90	84	88
1-pod smooth × 1-pod smooth *AaBb*　　　　*AaBb*	447	158	146	52
1-pod smooth × 3-pod smooth *AaBb*　　　　*aaBb*	318	98	323	104
1-pod smooth × 1-pod wrinkled *AaBb*　　　　*Aabb*	110	113	33	38
1-pod smooth × 3-pod smooth *AABb*　　　　*aaBb*	362	118	0	0

16. a.   wrinkled, green, white, short (short plants with white flowers grown from wrinkled, green seeds), because the genotype needed for a testcross is homozygous recessive.

    b.   For plant 1 all the testcross progeny would have green seeds because the plant is homozygous recessive for this characteristic. The distribution of the other characteristics in the testcross progeny would reflect the gametes produced by independent assortment of the three heterozygous genotypes:

$Rr \times rr$ $Pp \times pp$ $Tt \times tt$

                      1/2 P

          1/2 R

                      1/2 p

                      1/2 P

          1/2 r

                      1/2 p

- ↗ 1/2 T = 1/8 round purple tall
- ↘ 1/2 t = 1/8 round purple short
- ↗ 1/2 T = 1/8 round white tall
- ↘ 1/2 t = 1/8 round white short
- ↗ 1/2 T = 1/8 wrinkled purple tall
- ↘ 1/2 t = 1/8 wrinkled purple short
- ↗ 1/2 T = 1/8 wrinkled white tall
- ↘ 1/2 t = 1/8 wrinkled white short

    c.   For plant 2, all the testcross progeny would be short from yellow seeds because the plant is homozygous for these characteristics. The distribution of the other characteristics in the testcross progeny would reflect the gametes produced by independent assortment of the two heterozygous genotypes:

$Rr \times rr$ $Pp \times pp$

          1/2 R

          1/2 r

- ↗ 1/2 P = 1/4 round purple
- ↘ 1/2 p = 1/4 round white
- ↗ 1/2 P = 1/4 wrinkled purple
- ↘ 1/2 p = 1/4 wrinkled white

17. a.

	$Rr \times Rr$	$YY \times yy$	$Pp \times Pp$	$Tt \times tt$
1/32:	rr	Yy	pp	tt
	1/4 ×	1 ×	1/4 ×	1/2

    b.   1/32, the same as in part a, because Yy is the only possible outcome for this characteristic from this cross, and only the homozygous recessives will show the other three traits.

    c.

	$Rr \times Rr$	$YY \times yy$	$Pp \times Pp$	$Tt \times tt$
9/32:	R–	Y–	P–	T–
	3/4 ×	1 ×	3/4 ×	1/2

   d. 2/4 or 1/2, because either *RR* or *rr* will breed true for seed shape: 1/4 + 1/4 = 1/2

      Note that this is an example of the sum rule. The question asked for pure breeding for the characteristic and either homozygous round *or* wrinkled (which must be homozygous) will breed true.

   e. 1/16:   *RR*   *PP*

             1/4 × 1/4   (product rule)

   f. none, because all the progeny are *Yy* for seed color, a genotype that will not breed true

   g. 2/3, by conditional probability. The proportion of all the progeny with the genotype *RrYyppTt* is 1/16, but we are interested only in those progeny that have the round yellow white tall phenotype (3/32). So 1/16 ÷ 3/32 = 32/48 = 2/3.

18. a. $(3/4)^3$ = 27/64 = 42.2%

    b. The probability of a taster girl is 3/4 × 1/2 = 3/8. The probability of 3 taster girls is $(3/8)^3$ = 27/512 = 5.3%.

    c. 3/4 × 3/4 × 3/4 × 1/4 × 1/4 = 27/1024 = 2.6%

    d. 10 ways. The answer comes from the coefficient of the binomial: *n* = 5 children, *r* = 3 tasters, and *s* = 2 nontasters. Thus, (5!/3!2!) = 10.

    e. 270/1024 = 26.4%, by multiplying the coefficient from part d by the exponent of the binomial in part c. The exponent is: $(3/4)^3 (1/4)^2$.

19.

HYBRID SELF-FERTILIZED	OFFSPRING AND PHENOTYPIC RATIOS							
	A–B–D–	*A–B–dd*	*A–bbD–*	*A–bbdd*	*aaB–D–*	*aaB–dd*	*aabbD–*	*aabbdd*
AABBDd	3/4	1/4						
AaBBDd	9/16	3/16			3/16	1/16		
AaBBDD	3/16				1/16			
AaBbDD	9/16		3/16		3/16		1/16	
AaBbDd	27/64	9/64	9/64	3/64	9/64	3/64	3/64	1/64

20.

GENOTYPE OF FEMALE PARENT	GENOTYPE OF MALE PARENT	RATIOS OF PHENOTYPES IN PROGENY							
		ABD	*ABd*	*AbD*	*Abd*	*aBD*	*aBd*	*abD*	*abd*
AaBbDd	*AABbDd*	9	3	3	1				
aaBbDd	*Aabbdd*	1	1	1	1	1	1	1	1
aaBbDd	*AaBbdd*	3	3	1	1	3	3	1	1
Aabbdd	*AaBbDd*	3	3	3	3	1	1	1	1
Aabbdd	*AaBbdd*	1		1		1		1	

21. Here, A = gray, a = brown, B = spots, and b = no spots.

CROSSES PERFORMED				PROGENY				
				Phenotypes and Numbers Observed (NOT RATIOS)				
Male Parents		Female Parents		Gray with Spots	Gray without Spots	Brown with Spots	Brown without Spots	TOTAL Number
Phenotype	Genotype	Phenotype	Genotype					
Gray with spots	*AABB*	Gray with spots	*AABB*	180	0	0	0	180
Gray without spots	*Aabb*	Brown with spots	*aaBb*	52	47	46	54	199
Gray with spots	*AaBB*	Brown with spots	*aaBB*	101	0	101	0	202
Gray with spots	*AABb*	Brown with spots	*aaBb*	147	49	0	0	196
Brown with spots	*aaBb*	Brown with spots	*aaBb*	0	0	151	48	199
Brown with spots	*aaBb*	Gray with spots	*AaBb*	70	24	80	26	200
Gray with spots	*AaBb*	Gray with spots	*AaBb*	115	40	36	13	204

2
Mitosis and Meiosis

Problems

1. In an animal zygote with a complement of two homologous chromosome pairs, symbolized as *A, a* and *B, b,* which of the following is expected in its somatic cells during growth? Explain.
   a. *AaBB*
   b. *AABb*
   c. *AABB*
   d. *aabb*
   e. more than one of the above
   f. none of the above

2. If the individual in Problem 1 becomes an adult, which of the following combinations of chromosomes would you expect to find in its gametes? Explain.
   a. *Aa, AA, aa, Bb, BB, bb*
   b. *Aa, Bb*
   c. *A, a, B, b*
   d. *AB, Ab, aB, ab*

3. If one ignores the possibility of crossing-over during meiosis, what is the probability that a woman has received one member of each pair of homologous chromosomes present in her maternal grandmother?

4. Determine the diploid numbers of the following organisms from the per cell observations given below:
   a. 16 tetrads at metaphase I
   b. 60 chromatids at metaphase I
   c. 12 centromeres at anaphase I
   d. 12 centromeres at anaphase II
   e. 18 chromosomes at anaphase of mitosis

5. **MATCHING TEST** Choices may be used more than once. In the somatic cells of a certain animal species, 30 pairs of homologous chromosomes are present. How many chromosomes will be found in the following cells:

_____ mature egg	a. 15
_____ secondary oocyte	b. 30
_____ first polar body	c. 45
_____ brain cell	d. 60
_____ primary spermatocyte	e. 120
_____ spermatid	

6. The haploid amount of DNA in humans is 1.91×10^{12} daltons (1 dalton = 1.67×10^{-24} grams). How much DNA would be found in a male cell:
   a. during G_1?
   b. after the S phase?
   c. after mitotic telophase and cytokinesis?
   d. after telophase and cytokinesis of meiosis I?
   e. after telophase and cytokinesis of meiosis II during spermatogenesis?

7. If one pair of chromosomes contains genes *A* and *a*, and another pair of chromosomes contains genes *B* and *b*, with *A* dominant to *a* and *B* dominant to *b*, what is the probability of obtaining:
   a. an *AB* gamete from an *AaBb* individual?
   b. an *AB* gamete from an *AABb* individual?
   c. an *AABB* zygote from a cross *AaBb* × *AaBb*?
   d. an *AABB* zygote from a cross *aabb* × *AABB*?
   e. an *A–B–* phenotype from a cross *AaBb* × *AaBb*?
   f. an *A–B–* phenotype from a cross *aabb* × *AABB*?
   g. an *aaB–* phenotype from a cross *AaBb* × *AaBB*?

8. What types and frequencies of gametes are formed by the following genotypes if each pair of genes is found on a different pair of chromosomes?
   a. *AaBBCc*
   b. *DdEEffGg*
   c. *MmNnOo*
   d. *AaBbCcDd*

9. Human somatic cells have 46 chromosomes. How many chromosomes are in the following:
   a. mature egg
   b. first polar body
   c. sperm cell
   d. spermatid
   e. primary spermatocyte
   f. brain cell
   g. secondary oocyte
   h. spermatogonium

10. Draw the alignment of two pairs of chromosomes labeled A, A' for the first pair and B, B' for the second pair at mitotic anaphase, meiotic anaphase I, and meiotic anaphase II.

11. Indicate whether each statement below is true or false.
    a. A diploid organism with $1n = 4$ can produce 16 different kinds of gametes in relation to numbers of maternal and paternal kinds of chromosomes.
    b. A diploid organism with $1n = 3$ has 1 chance in 8 of producing a gamete with all paternal chromosomes.
    c. Even though only 1 functional egg results from each oogenesis, eggs are as genetically variable as sperm.
    d. A single $2n$ cell undergoing spermatogenesis will produce 4 genetically distinct types of gametes if no crossing-over occurs.
    e. The random alignment of different tetrads at metaphase I of meiosis accounts for the law of segregation.
    f. Homologous chromosomes in humans normally have identical loci with identical alleles.
    g. The diploid number of chromosomes is double the haploid number.

    h. A somatic cell that will not divide again will remain in the G_1 stage of interphase.

    i. The correct sequence of events during prophase I of meiosis is: synapsis → bivalent formation → tetrad formation → chiasma formation.

    j. A secondary oocyte has half as many chromosomes as a primary oocyte.

    k. In a primary spermatocyte containing 30 chromosomes, 15 of the chromosomes are paternal.

    l. Separation of sister chromatids occurs during meiotic anaphase I and is the basis for the law of segregation.

    m. In mice ($2n = 40$), 20 tetrads will be present at metaphase of mitosis.

    n. In mice ($2n = 40$), 40 linkage groups will be present.

    o. In mice ($2n = 40$), 2^{40} different combinations of chromosomes may be seen in a large group of sperm cells if the effects of crossing-over are ignored.

12. In each case, circle the correct choice(s).

    a. If an individual has the genotype *AaBBDdEeFfGghh*, how many types of gametes can be produced, assuming each gene pair is in a different linkage group and crossing-over does not occur?

        2   4   8   14   16   32   64 2^7 7^2 $(1/2)^5$

    b. In the individual in a above, which of the following chromosome complements may be found in one of his sperm cells?

        ADEFh   *ABDEFGH*   *aBdEfGh*   *AaBBDdEeFfGghh*   none of these

    c. If a zygote receives the three chromosome pairs with A and a, B and b, and D and d, which of the following chromosome complements will be found in its skin cells?

        AAaaBBbbDDdd   *AABBDD*   *aabbdd*   *AaBbDd*   *ABD*   *abd*

13. If a statement is true, do not change it. If a statement is false, change it so that it becomes true.

    a. In meiosis, the chromosomal material doubles during the last premeiotic interphase.

    b. The greater the number of chromosome pairs, the greater is the effect of meiosis in producing genetically distinct gametes.

    c. The diploid number of chromosomes is half the haploid number.

    d. A cell that will not divide again remains in the G_2 stage of interphase.

    e. The correct sequence of events during meiosis is: synapsis → tetrad formation → bivalent formation → chiasma formation.

    f. An organism with $n = 3$ has 1 chance in 8 of producing a gamete with all maternally derived chromosomes.

    g. An organism with $n = 3$ can produce 8 different kinds of gametes in relation to combinations of maternal and paternal chromosomes.

    h. Separation of sister chromatids occurs during anaphase I of meiosis and is the physical basis for the law of independent assortment.

    i. A single diploid cell that undergoes meiosis in the male testis will produce 4 genetically different gametes if crossing-over is ignored.

    j. The independent alignment of different tetrads at metaphase I of meiosis is the physical basis of the law of independent assortment.

    k. If a diploid cell has 20 pairs of chromosomes, there will be 40 linkage groups present.

    l. If a diploid cell has 20 pairs of chromosomes, there will be 10 tetrads during prophase I of meiosis.

   m. If a diploid cell has 20 pairs of chromosomes, there will be 20 bivalents during prophase I of meiosis.

   n. If a diploid cell has 20 pairs of chromosomes, there will be 80 chromatids present during metaphase of mitosis.

   o. If a diploid cell has 20 pairs of chromosomes, the organism can make $(20)^2$ types of gametes if crossing-over is ignored.

Challenge Problems

14. For an organism with a diploid number of 30 chromosomes, determine the per cell number of the following:
   a. chromatids at metaphase of mitosis
   b. chromosomes at metaphase of mitosis
   c. centromeres at metaphase of mitosis
   d. chromatids at anaphase of mitosis
   e. chromosomes at anaphase of mitosis
   f. centromeres at anaphase of mitosis
   g. tetrads at metaphase I of meiosis
   h. chromosomes at metaphase I of meiosis
   i. chromatids at metaphase I of meiosis
   j. chromosomes at anaphase I of meiosis
   k. chromatids at anaphase I of meiosis
   l. tetrads at metaphase II of meiosis
   m. chromosomes at metaphase II of meiosis
   n. chromosomes at anaphase II of meiosis

15. A certain animal has 4 pairs of homologous chromosomes designated *A* and *a* for pair 1, *B* and *b* for pair 2, *D* and *d* for pair 3, and *E* and *e* for pair 4. The following questions pertain to this individual:
   a. Which chromosomes could *not* have been contributed as a set by his father?
     a) *A, B, D, E*   b) *A, b, D, E*   c) *B, d*   d) *A*   e) *E, e*
   b. How many chromosomes came from his father?
     a) 2   b) 4   c) 6   d) 8   e) none of these
   c. Which chromosome complement is found in his skin cells?
     a) *aabbDDEE*   b) *AaBbDdEe*   c) *AAee*   d) *ABDE*   e) *AbDe*
     f) can't tell for sure because it will vary
   d. Which complement could be found in his sperm cells?
     a) *AABBDDEE*   b) *AaBbDdEe*   c) *AAee*   d) *ABDE*
     e) none of these is possible
   e. How many chromosomes will he pass on to his daughter?
     a) 2   b) 4   c) 6   d) 8   e) a variable number
   f. How many chromatids are present at metaphase I in his primary spermatocytes?
     a) 0   b) 4   c) 8   d) 12   e) 16
   g. How many pairs of homologous chromosomes are present at metaphase II in his secondary spermatocytes?
     a) 0   b) 4   c) 8   d) 12   e) 16

h. How many different combinations will be present in a large sample of his sperm cells (ignoring the effects of crossing-over)?

    a) 2^3 b) 3^2 c) 2^4 d) 4^2 e) 3^4 f) 4^3

16. Where possible (specify if it is not possible), give the stages of mitosis and meiosis I or II at which:

STAGE OF MITOSIS	STAGE OF MEIOSIS	
a. _____	_____	Homologous replicated chromosomes move toward opposite poles.
b. _____	_____	Nuclear membranes re-form.
c. _____	_____	Centromeres align at the equator in diploid cells.
d. _____	_____	Homologous chromosomes pair up.
e. _____	_____	Microtubule proteins are synthesized.
f. _____	_____	Chromosomes are replicating.
g. _____	_____	Sister chromatids become unreplicated chromosomes.
h. _____	_____	Reciprocal exchange of nonsister chromatids occurs.
i. _____	_____	Unpaired chromosomes congregate at the equator of the spindle.

Team Problems

17. Explain why each of the following statements is false.
    a. Skin cells and gametes of the same animal contain the same number of chromosomes.
    b. Any chromosome may pair with any other chromosome in the same cell during meiosis.
    c. The gametes of an animal may contain more maternal chromosomes than its body cells contain.
    d. A cell that is not going to divide again will remain in the G_2 portion of interphase.
    e. A cell at the end of interphase G_2 has a $2c$ amount of DNA.
    f. Mendel's law of segregation is illustrated by the movement of homologous chromosomes during metaphase I of meiosis.
    g. Segregation of sister chromatids occurs during anaphase I of meiosis and is the physical basis of the law of segregation.
    h. The independent movement of individual homologous chromosomes at metaphase I of meiosis is the physical basis for the law of independent assortment.
    i. In the testis, a single diploid cell that undergoes meiosis will produce four genetically different gametes if crossing-over is ignored.

18. Two varieties of plants have 6 chromosomes each in their diploid cells. In variety 1, the first pair of chromosomes has "knobs" at one end. In variety 2, the second pair of chromosomes has "satellites" at one end, and the third pair has a small section of each chromosome missing (a "deletion"). These chromosomes are diagrammed below.

variety 1   *variety 2*

Variety 1 was mated with variety 2. From the possible answers given, choose the best choice (or choices) for the following questions:

a. types of gametes produced by the variety 1 plant

b. types of gametes produced by the variety 2 plant

c. chromosome constitution of the F_1 plants produced by the variety 1 × variety 2 cross

d. types of gametes produced by the F_1 plants

e. all choices indicating normal diploid cells

f. all choices indicating gametes or zygotes that would be formed if Mendel's laws (as they apply to chromosome movement during meiosis) were violated

g. If the F_1 plant is backcrossed to the variety 1 plant, what is the chance of producing an offspring with the same chromosome constitution as the variety 1 plant?

h. If two F_1 plants mate, what is the chance of producing an offspring with the same chromosome constitution as the variety 1 plant?

Possible Answers

19. Assume that a certain organism has a diploid number of 2 pairs of chromosomes ($2n = 4$) in its body cells and that these chromosomes may be referred to as the A, a pair and the B, b pair. Thus, a normal diploid cell in this organism has one copy each of the $A, a, B,$ and b chromosomes. Identify, from the choices given below, the stage of mitotic or meiotic cell division for each diagram.

Choices: Each may be used more than once or not at all.

a. interphase

b. anaphase of mitosis

c. prophase of mitosis

d. metaphase of mitosis

e. telophase of mitosis

f. metaphase I of meiosis

g. early prophase I of meiosis (synapsis and bivalents)

h. mid–prophase I of meiosis (tetrads form)

i. late prophase I of meiosis (chiasma present)

j. telophase I of meiosis

k. anaphase I of meiosis

l. telophase II of meiosis

m. prophase II of meiosis

n. anaphase II of meiosis

o. metaphase II of meiosis

p. none of the above (an impossible situation if Mendel's laws are followed)

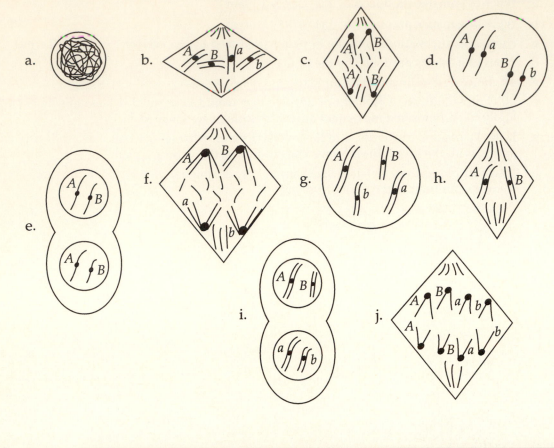

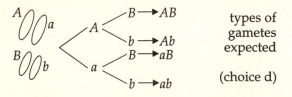

Solutions

1. Somatic cells are produced by mitosis. Mitosis faithfully reproduces and distributes whatever chromosomes are present in the original cell so that all subsequent cells have identical sets of chromosomes. Because the zygote had 4 different chromosomes, *A*, *a*, *B*, and *b*, all somatic cells would likewise be *AaBb*. Thus, none of the choices is correct.

2. Gametes are produced by meiosis in animals. Meiosis involves segregation and independent assortment of homologous chromosome pairs. Thus, if the diploid gonadal cell has *A*, *a*, *B*, and *b* chromosomes,

   types of gametes expected

   (choice d)

3. 1/8,388,608. For each pair of homologs, the woman's mother has one chromosome from her mother and one from her father. Similarly, her mother received one of her chromosomes from the woman's grandmother and one from the woman's grandfather. The chance the woman's mother passes on a chromosome from the grandmother is, thus, 1/2 for each of the 23 pairs, or $(1/2)^{23} = 1/8,388,608$.

4. a. 32. Tetrads are formed by synapsis of homologs.

    b. 30. There are two sister chromatids for each chromosome.

    c. 12. Each centromere corresponds to a chromosome, and there is no division of the centromeres at anaphase I.

    d. 12. Formation of daughter chromosomes from sister chromatids has occurred, and the haploid number of 6 is moving toward each pole.

    e. 9. Formation of daughter chromosomes from sister chromatids has occurred, and each diploid set of 9 is moving toward the opposite pole.

5. Here, $2n = 60$ chromosomes (30 pairs).

    A mature ($1n$) egg has 30 unreplicated chromosomes.

    A secondary oocyte ($1n$) has 30 replicated chromosomes.

    A first polar body ($1n$) has 30 replicated chromosomes.

    A brain cell ($2n$) has 60 unreplicated chromosomes (in G_1).

    A primary spermatocyte has 60 replicated chromosomes.

    A spermatid has 30 unreplicated chromosomes.

6. a. 3.82×10^{12} daltons, twice the haploid amount ($2c$)

    b. 7.64×10^{12} daltons, the DNA has replicated ($4c$)

    c. 3.82×10^{12} daltons, the cell is in G_1 ($2c$)

    d. 3.82×10^{12} daltons, although the cell is haploid, each chromosome possesses 2 chromatids

    e. 1.91×10^{12} daltons, the haploid c amount because each sister chromatid has become a daughter chromosome incorporated into a cell

7. a. 1/4, because 1/2 (A) × 1/2 (B) = 1/4 AB

    b. 1/2, because 1 (A) × 1/2 (B) = 1/2 AB

    c. 1/16

    d. 0

    e. 9/16

    f. all

    g. 1/4

8. a. ABC, ABc, aBC, aBc

    b. $DEfG, DEfg, dEfG, dEfg$

    c. $MNO, MNo, MnO, mNO, Mno, mnO, mNo, mno$

    d. $ABCD, ABCd, ABcD, AbCD, aBCD, abCD, aBcD, aBCd, AbcD, AbCd, ABcd, Abcd, aBcd, abCd,$
        $abcD, abcd$

9. a. 23
    b. 23
    c. 23
    d. 23
    e. 46
    f. 46
    g. 23
    h. 46

10.

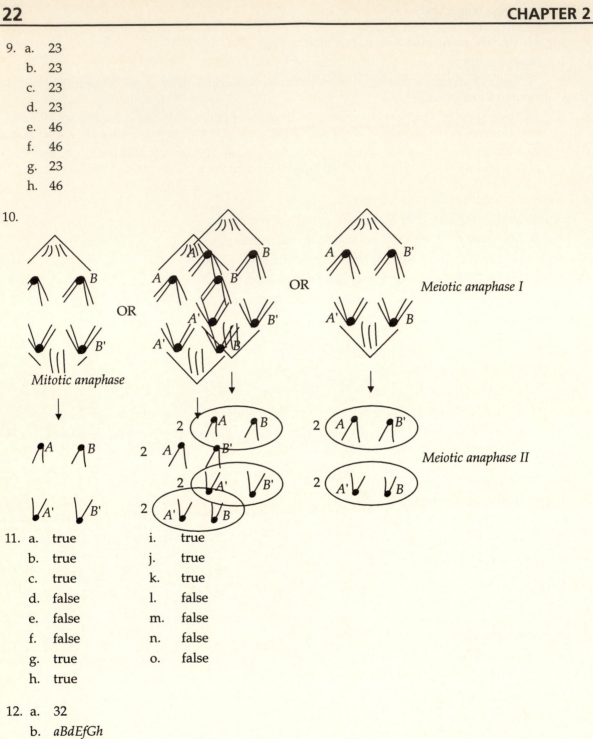

Mitotic anaphase

OR

OR

Meiotic anaphase I

Meiotic anaphase II

11. a. true   i. true
    b. true   j. true
    c. true   k. true
    d. false   l. false
    e. false   m. false
    f. false   n. false
    g. true   o. false
    h. true

12. a. 32
    b. *aBdEfGh*
    c. *AaBbDd*

13. Statements a, b, f, g, j, m, and n are true.

    c. The diploid number of chromosomes is **twice** the haploid number.
    d. A cell that will not divide again remains in the **G₁** or **G₀** stage of interphase.
    e. The correct sequence of events during meiosis is: synapsis → **bivalent** formation → **tetrad** formation → **chiasma** formation.

h. Separation of **homologous chromosomes** occurs during anaphase I of meiosis and is the physical basis for the law of independent assortment.

i. A single diploid cell that undergoes meiosis in the male testis will produce **2** genetically different gametes if crossing-over is ignored.

k. If a diploid cell has 20 pairs of chromosomes, there will be **20** linkage groups present.

l. If a diploid cell has 20 pairs of chromosomes, there will be **20** tetrads during prophase I of meiosis.

o. If a diploid cell has 20 pairs of chromosomes, the organism can make $(2)^{20}$ types of gametes if crossing-over is ignored.

14. a. 60, two sister chromatids for each chromosome

   b. 30, each chromosome still has two sister chromatids

   c. 30, the same as the number of chromosomes

   d. 0, the sister chromatids have now become daughter chromosomes after duplication and division of the centromere

   e. 60, one daughter chromosome is moving toward each pole and will be incorporated into separate cells during telophase and cytokinesis

   f. 60, the same as the number of chromosomes

   g. 15, one for each pair of homologous chromosomes

   h. 30, each chromosome still has two sister chromatids

   i. 60, two sister chromatids for each chromosome

   j. 30, there is no formation of daughter chromosomes during the first division of meiosis, one chromosome is moving toward each pole and will be incorporated into separate cells during telophase I and cytokinesis

   k. 60, two sister chromatids for each chromosome

   l. 0, no tetrads are formed, the cells are haploid, and there are no homologous pairs of chromosomes

   m. 15, the haploid number, each, however, still has two sister chromatids as a result of the S stage, which preceded meiosis I

   n. 30, as a result of formation of daughter chromosomes from duplication and division of the centromere (One copy of each is moving toward each pole and will be incorporated into separate cells after telophase II and cytokinesis.)

15. Here, $2n = 8$ (4 pairs). So

   a. Choice e is correct because both *E* and *e* could not have come from the father (one member of each homologous pair comes from each parent).

   b. Half come from the father (choice b = 4).

   c. Skin cells contain the same number and kinds of chromosomes as were found in the zygote (choice b = *AaBbDdEe*).

   d. Sperm cells contain a haploid combination of one chromosome from each homologous pair (choice d = *ABDE*).

   e. He will pass half his chromosomes (one of each pair) to his daughter (choice b = 4).

   f. During first meiotic division, the chromosomes are duplicated structures, each with two chromatids (choice e = 16 chromatids).

   g. At metaphase II, no homologous pairs are present, just one member of *each* homologous pair.

   h. Choice c is correct: $2^x = 2^4 = 16$ different combinations of chromosomes.

16.

	STAGE OF MITOSIS	STAGE OF MEIOSIS
a.	not possible	anaphase I
b.	telophase	telophase II (and sometimes telophase I)
c.	metaphase	metaphase I
d.	not possible	prophase I
e.	interphase G_2	interphase G_2
f.	interphase S	interphase S
g.	anaphase	anaphase II
h.	not possible	prophase I
i.	metaphase	metaphase II

17. a. Skin cells are diploid (2n), whereas gametes are haploid (1n).

b. Only homologous chromosomes pair during meiosis. Sex chromosomes, although different, will also pair because they have a small region of genetic homology between them.

c. This is impossible because, even if a gamete had *all* maternal chromosomes, their number could not exceed the number found in somatic cells. One chromosome of each homologous pair is maternal (from the egg) in origin, the other being paternal (from the sperm). Thus, half of the diploid number of chromosomes is maternal in origin.

d. A nondividing cell remains in G_1 of interphase, a stage in which the chromosomes are unreplicated and generalized cell metabolism occurs.

e. A cell at the end of G_2 has replicated chromosomes and has a 4c amount of DNA.

f. The law of independent assortment is illustrated by the independent alignment of tetrads at the equatorial plate during metaphase I. Segregation occurs during anaphase I when homologous chromosomes move toward opposite poles of the spindle.

g. Segregation of homologous chromosomes occurs during anaphase I of meiosis. Segregation of sister chromatids occurs during anaphase II of meiosis.

h. Individual homologous chromosomes do not move to the equatorial plate independent of each other at metaphase I of meiosis. They are synapsed and align as tetrad units at metaphase I.

i. A single diploid cell will produce four sperm cells of two chromosomal combinations only. In order to see the full array of gamete types due to independent assortment, a large sample of sperm cells must be observed.

18. Here, 2n = 6 (3 pairs).

a. choice K

b. choice J

c. choice E

d. choices G through N

e. choices A, B, C, E, and F

f. choices D, O, P, and Q

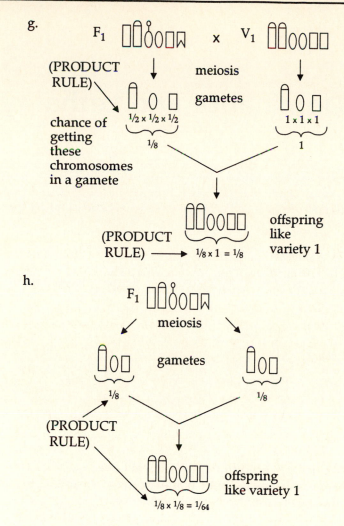

g.

(PRODUCT RULE)

chance of getting these chromosomes in a gamete

meiosis

gametes

½ x ½ x ½

⅛

1 x 1 x 1

1

(PRODUCT RULE) ⟶ ⅛ x 1 = ⅛

offspring like variety 1

h.

F_1

meiosis

gametes

⅛

⅛

(PRODUCT RULE)

⅛ x ⅛ = ¹⁄₆₄

offspring like variety 1

19. a. choice A
    b. choice D
    c. choice N
    d. choice P
    e. choice L
    f. choice K
    g. choice C
    h. choice M
    i. choice J
    j   choice B

3
Dominance and Lethality

Problems

1. Two curly-winged *Drosophila* are mated and produce an F_1 that consists of 341 curly and 162 normal flies. Explain.

2. In plants, self-sterility may occur when pollen with a particular allele cannot facilitate fertilization of a plant carrying an identical allele. Consider a plant species with a series of 6 self-sterility alleles (S_1 to S_6). For each cross below, indicate whether it is sterile, partially fertile, or completely fertile and give the expected progeny.

	POLLEN PARENT	FEMALE PARENT
a.	S_1S_3	S_1S_3
b.	S_1S_4	S_1S_3
c.	S_4S_5	S_3S_4
d.	S_2S_6	S_3S_4

3. In cattle, the allele producing red coat color (R) is incompletely dominant over the allele (r) for white coat color. Heterozygotes (Rr) are roan colored. Cattle homozygous for a different gene, "short spine" (aa), are deformed and die shortly after birth. Heterozygous carriers (Aa) are completely normal. If a series of matings is conducted between Aa roan animals, what will be the phenotypic ratio:

   a. at birth?

   b. several days after birth?

4. Two mice with yellow fur mate and produce several litters of offspring. The F_1 consists of 34 yellow and 16 black mice. Explain.

5. In snapdragon plants, the allele for red flower color has an effect that is incompletely dominant over the effect of the white color allele. If a cross between two plants produced 18 red, 32 pink, and 15 white plants, what are the phenotypes of the parents?

6. What ratios of flower color in snapdragons are expected among the offspring of the following crosses:

   a. red × red

   b. red × pink

   c. white × pink

   d. pink × pink

7. In cattle, the effect of the allele producing red coat color (*R*) is incompletely dominant over the effect of the allele producing white coat color (*r*), the *Rr* heterozygote being roan colored. In addition, the effects of alleles for the absence of horns show complete dominance: *HH* and *Hh* are hornless ("polled") and *hh* is horned. Genes at the *R* and *H* loci show independent assortment.

   a. Give the phenotype of the F_1 offspring of *RRHH* × *rrhh*.
   b. Give the expected phenotypes and their proportions among the offspring of an F_1 × F_1 mating.
   c. Give the expected phenotypes and their proportions among the offspring of an F_1 × the white-horned parent.

8. Consider the following data from matings between pheasants differing in plumage color:

		PROGENY		
MALES	FEMALES	LIGHT	BUFF	RING
light	light	92	0	0
light	buff	78	70	0
buff	light	14	12	0
light	ring	0	208	0
buff	buff	75	141	68
buff	ring	0	112	128
ring	buff	0	34	38
ring	ring	0	0	347

   a. Using letters to designate genes, propose a hypothesis that best explains these results.
   b. On the basis of the letters you choose, give the genotypes of the light, ring, and buff genotypes.

Challenge Problems

9. In certain bird species, crosses were preformed to determine the inheritance of tail feather color and length. The parents and offspring of three such crosses are given below. Based on the results of these crosses, fill in the color and length phenotypes below, determine which color and length are dominant, and give the genotypes for each bird in the three matings.

   Fill in the phenotypes:

COLOR GENOTYPES
AA: _____
Aa: _____
aa: _____
Dominant color: _____

LENGTH GENOTYPES:
BB: _____
Bb: _____
bb: _____
Dominant length: _____

Give all of the following genotypes:

CROSS 1

P: yellow short × blue long
     feathers   feathers

_____   ↓   _____

F$_1$:   39 green short _____

       43 green long _____

CROSS 2

     yellow long × green short
     feathers   feathers

_____   ↓   _____

19 green short _____

22 yellow short _____

20 green long _____

18 yellow long _____

CROSS 3

P: green short feathers × green short feathers

_____   ↓   _____

F$_1$:   13 blue short _____   28 green short _____

       15 yellow short _____   7 blue long _____

       11 green long _____   5 yellow long _____

10. In mice, nude is a recessive trait (gene *n*) in which newborn mice lack fur and die within a week of birth. Dominant allele *N* allows normal fur to develop, a nonlethal condition. In addition, an independently assorting dominant trait is yellow fur color, caused by gene A^Y, which also acts as a recessive lethal allele during early embryonic development. The recessive allele for black fur color (gene *a*) does not have a lethal effect.

   a. The following information is known about four mice strains. Give their most likely genotypes:

	MOUSE INFORMATION	GENOTYPE
I	black fur, has not produced any nude mice	_____
II	black fur, has produced some nude mice	_____
III	yellow fur, has not produced any nude mice	_____
IV	yellow fur, has produced some nude mice	_____

   b. In the following crosses, indicate the expected potential phenotypes and ratios of the offspring at the indicated stages of development:

   strain II ♂ × strain IV ♀

offspring phenotypes	ratio at fertilization	ratio at birth	ratio at age 21 days
_____	_____	_____	_____
_____	_____	_____	_____
_____	_____	_____	_____
_____	_____	_____	_____

strain IV ♂ × strain IV ♀

offspring phenotypes	ratio at fertilization	ratio at birth	ratio at age 21 days
_____	_____	_____	_____
_____	_____	_____	_____
_____	_____	_____	_____
_____	_____	_____	_____

Team Problems

11. Determine the genotypic and phenotypic ratios of offspring at birth resulting from each of the following dihybrid crosses (assume that lethal alleles exert their effects early in embryonic development):

		GENE CHARACTERISTICS	
	PARENTAL GENOTYPES	FIRST PAIR	SECOND PAIR
a.	$AaBb \times AaBb$	complete dominance	complete dominance
b.	$Aab_1b_2 \times Aab_1b_2$	complete dominance	incomplete dominance
c.	$a_1a_2b_1b_2 \times a_1a_2b_1b_2$	incomplete dominance	incomplete dominance
d.	$AaBb \times AaBb$	complete dominance	recessive lethal
e.	$a_1a_2Bb \times A_1A_2Bb$	incomplete dominance	recessive lethal
f.	$AaBb \times AaBb$	recessive lethal	recessive lethal

12. a. In a certain species of plant, the fruits show variation in both shape and color. Genes at the (A, a) locus control the shape of the fruit, and genes at the (B, b) locus control the color of the fruit. A cross between two doubly heterozygous plants $(AaBb \times AaBb)$ produces viable offspring in the total numbers given below. Give the exact genotypes of all offspring.

$AaBb$ × $AaBb$
↓

	EXACT GENOTYPES
105 with long, red fruit	_____
203 with round, red fruit	_____
397 with round, brown fruit	_____
196 with long, brown fruit	_____
204 with round, yellow fruit	_____
99 with long, yellow fruit	_____

b. Give the phenotypes for each of the following genotypes:

SHAPE GENOTYPES	COLOR GENOTYPES
_AA:_ _____	_BB:_ _____
_Aa:_ _____	_Bb:_ _____
_aa:_ _____	_bb:_ _____

Solutions

1. The dominant gene for curly wings acts as a recessive lethal (C/C homozygotes die due to recessive lethality; $C/+$ heterozygotes show both the dominant phenotypes: viability and curly wings). Matings between curly flies must be between heterozygotes, and the phenotypic ratio of the progeny will be 2/3 curly: 1/3 normal because homozygous curly embryos fail to survive.

2. a. none (completely infertile)
   b. S_1S_4, S_3S_4 (partially fertile)
   c. S_3S_5, S_4S_5 (partially fertile)
   d. S_2S_3, S_2S_4, S_3S_6, S_4S_6 (completely fertile)

3. a. The cross is $AaRr \times AaRr$, in which aa is recessive lethal and Rr is incompletely dominant roan color. The expected ratio is 3 red normal: 6 roan normal: 3 white normal: 1 red deformed: 2 roan deformed: 1 white deformed.
   b. 2 red: 4 roan: 2 white (a 1:2:1 ratio). This is a modification of the above ratio after the death of the deformed cattle.

4. Yellow fur is the heterozygous expression of a recessive lethal gene; 34:16 is a close approximation of a 2:1 ratio.

5. $Rr \times Rr$ (both pink)

6. a. all red
   b. 1 red: 1 pink
   c. 1 white: 1 pink
   d. 1 red: 2 pink: 1 white

7. a. roan hornless
   b. 1 red horned: 2 roan horned: 1 white horned: 3 red hornless: 6 roan hornless: 3 white hornless
   c. 1 roan horned: 1 roan hornless: 1 white hornless: 1 white horned

8. LL = light, Ll = buff, ll = ring; L and l show incomplete dominance

9. Hint: consider the ratios of progeny separately for each trait.

COLOR GENOTYPES	LENGTH GENOTYPES
AA: **yellow**	BB: **lethal**
Aa: **green**	Bb: **short**
aa: **blue**	bb: **long**

Dominant color: **neither**

Dominant length: **short (also recessive lethal)**

CROSS 1

			CROSS 2	

P: yellow short × blue long   yellow long × green short
        feathers   feathers   feathers   feathers
        AABb   ↓ *aabb*   *AAbb*   ↓ *AaBb*

F_1:   39 green short   *AaBb*   19 green short   *AaBb*
        43 green long   *Aabb*   22 yellow short   *AABb*
                                         20 green long   *Aabb*
                                        18 yellow long   *AAbb*

CROSS 3

P: green short feathers × green short feathers

        AaBb   ↓ *AaBb*

F_1:   13 blue short   *aaBb*   28 green short   *AaBb*
        15 yellow short *AABb*   7 blue long   *aabb*
        11 green long   *Aabb*   5 yellow long   *AAbb*

10. a.

	MOUSE INFORMATION	GENOTYPE
I	black fur, has not produced any nude mice	*aaNN*
II	black fur, has produced some nude mice	*aaNn*
III	yellow fur, has not produced any nude mice	A^YaNN
IV	yellow fur, has produced some nude mice	A^YaNn

b. **strain II ♂ × strain IV ♀**

offspring phenotypes	ratio at fertilization	ratio at birth	ratio at age 21 days
yellow, normal fur:	3/8	3/8	3/6
black, normal fur:	3/8	3/8	3/6
yellow, nude:	1/8	1/8	0
black, nude:	1/8	1/8	0

strain IV ♂ × strain IV ♀

offspring phenotypes	ratio at fertilization	ratio at birth	ratio at age 21 days
yellow, normal fur:	3/8	3/8	3/6
black, normal fur:	3/16	3/12	3/9
yellow, nude:	3/16	2/12	0
black, nude:	1/16	1/12	0

11.

GENOTYPIC	PHENOTYPIC
a. 1 *AABB*: 2 *AaBB*: 2 *AABb*: 4 *AaBb*: 1 *AAbb*: 2 *Aabb*: 1 *aaBB*: 2 *aaBb*: 1 *aabb*	9 *A–B–*: 3 *A–bb*: 3 *aaB–*: 1 *aabb*
b. 1 AAb_1b_1: 2 Aab_1b_1: 1 aab_1b_1: 2 AAb_1b_2: 4 Aab_1b_2: 2 aab_1b_2: 1 AAb_2b_2: 2 Aab_2b_2: 1 aab_2b_2	3 $A–b_1b_1$: 6 $A–b_1b_2$: 3 $A–b_2b_2$: 1 aab_1b_1: 2 aab_1b_2: 1 aab_2b_2
c. 1 $a_1a_1b_1b_1$: 2 $a_1a_2b_1b_1$: 1 $a_2a_2b_1b_1$: 2 $a_1a_1b_1b_2$: 4 $a_1a_2b_1b_2$: 2 $a_2a_2b_1b_2$: 1 $a_1a_1b_2b_2$: 2 $a_1a_2b_2b_2$: 1 $a_2a_2b_2b_2$	same as the genotypes
d. 1 *AABB*: 2 *AaBB*: 2 *AABb*: 4 *AaBb*: 1 *aaBB*: 2 *aaBb* (*––bb* = lethal)	3 *A–B–*: 1 *aaB–*
e. 1 *AABB*: 2 *AaBB*: 2 *AABb*: 4 *AaBb*: 1 *aaBB*: 2 *aaBb* (*––bb* = lethal)	1 *AAB–*: 2 *AaB–*: 1 *aaB–*
f. 1 *AABB*: 2 *AaBB*: 2 *AABb*: 4 *AaBb*: (*aa––* and *––bb* = lethal)	all *A–B–*

12. Notice that there are 804 round and 400 long, a 2:1 ratio indicative of recessive lethal for homozygous round. The ratio of red: brown: yellow is 1:2:1, indicative of incomplete dominance.

a. **AaBb** × **AaBb**   **EXACT GENOTYPES**
↓

105 with long, red fruit	*aaBB*
203 with round, red fruit	*AaBB*
397 with round, brown fruit	*AABb*
196 with long, brown fruit	*aaBb*
204 with round, yellow fruit	*AAbb*
99 with long, yellow fruit	*aabb*

b.

AA:	lethal	*BB:*	red
Aa:	round	*Bb:*	brown
aa:	long	*bb:*	yellow

4
Multiple Alleles

Problems

1. In humans, thalassemia shows incomplete dominance; heterozygotes exhibit a mild form of the disease (thalassemia minor) and homozygotes, a much more severe form (thalassemia major). The inheritance of the ABO blood group system exhibits both complete dominance (alleles A and B are recessive to allele O) and codominance (alleles A and B). Both members of a couple have thalassemia minor. The husband has blood type A, and his mother was blood type O. The wife has blood type AB. What is the probability that they will have a child with:

    a. no anemia and blood type A?

    b. thalassemia minor and blood type B?

    c. thalassemia major and blood type AB?

2. In mice, a series of five alleles has been associated with fur pattern. In order of complete dominance, these alleles are: A^Y (homozygous lethal) for yellow fur > A^L, agouti with light belly > A^+ agouti > a^t black and tan > a, black. For each of the following crosses, give the coat colors of the parents and the phenotypic ratios expected among the offspring.

    a. $A^Y A^L \times A^Y A^+$

    b. $A^Y a \times A^L a^t$

    c. $a^t a \times A^L a^t$

    d. $A^Y a^t \times A^Y A^+$

3. In guinea pigs, one of the genes that affects coat color has a number of different alleles. In a certain strain of guinea pig, homozygous combinations of these alleles produce the following phenotypes: CC = black, $c^d c^d$ = cream, $c^k c^k$ = sepia, $c^a c^a$ = albino. These alleles show complete dominance in the order $C > c^k > c^d > c^a$. What are the expected phenotypes and their proportions among the offspring of the following crosses:

    a. homozygous black × homozygous sepia

    b. homozygous black × homozygous cream

    c. homozygous black × homozygous albino

    d. homozygous sepia × homozygous cream

    e. the $F_1 \times F_1$ of c

    f. the F_1 of a × the F_1 of d

    g. the F_1 of b × the F_1 of d

4. Give the genotypes of the following parents:

PHENOTYPES OF PARENTS	PROPORTIONS OF OFFSPRING			
	A	B	AB	O
a. B × B		3/4		1/4
b. B × AB		1/2	1/2	
c. B × A		1/2	1/2	
d. B × A	1/4	1/4	1/4	1/4
e. B × AB	1/4	1/2	1/4	
f. B × O		1/2		1/2

5. Which of the following males can be excluded as the possible father of a type O, Rh+, MN offspring when the maternal phenotype is type O, rh–, MN?
   a. AB Rh+ M
   b. A Rh+ MN
   c. B rh– MN
   d. O rh– N

6. If a series of four alleles exists in a given diploid (2*n*) species, how many alleles would be present in:
   a. a chromosome?
   b. a pair of chromosomes?
   c. an individual member of the species?
   How many genotypic combinations would occur in the entire species?

7. In a certain animal species, four alleles (c^+, c^1, c^2, and c) have their locus in chromosome I and another series of two alleles (d^+ and d) has its locus in chromosome II. How many different genotypes with respect to these two series of alleles are theoretically possible in the population?

Use the following information for the next three problems. A series of multiple alleles for fur color is reported in mice. One series of breeding data is:

PARENTS	F_1
plain black × white bellied	white bellied
plain black × dark bellied	dark bellied
white bellied × dark bellied	white bellied

8. The cross of two heterozygous dark-bellied animals will produce what phenotypes and ratios in the F_1?

9. A cross between two white-bellied animals produced an F_1 ratio of 3 white bellied: 1 dark bellied. What were the parental genotypes?

10. Later, an additional allele in this series, producing black and tan coat, was discovered. Crosses of black and tan with white bellied produced all white bellied in the F_1, and black and tan × dark bellied produced all dark bellied in the F_1. Crossing a different black and tan mouse with a plain black produced all black and tan F_1s. When these F_1s were interbred, a ratio of 3 black and tan: 1 plain black was produced in the offspring. What is the order of relative dominance among the four phenotypes?

11. In guinea pigs, one of the genes that affects coat color has a number of alleles that show complete dominance in the order: C, black > c^k, sepia > c^d, cream > c, albino. In the table below, give the most probable genotypes for the parents that produce the following offspring:

PHENOTYPES OF PARENTS	PHENOTYPES OF PROGENY			
	BLACK	SEPIA	CREAM	ALBINO
a. black × black	22	0	0	7
b. sepia × cream	0	24	11	12
c. black × albino	13	0	12	0
d. black × cream	19	20	0	0
e. black × sepia	14	8	6	0

In the following two problems on blood typing, the MN blood group system shows codominance, so genotype MM displays phenotype M, genotype NN = phenotype N, and genotype MN = phenotype MN.

12. In each of the following cases of disputed paternity, indicate whether, on the basis of the genetic evidence, the accused male should be eliminated from further consideration as the father.

MOTHER	CHILD	ACCUSED MALE	ELIMINATE MALE?
A Rh+	O Rh+	AB Rh+	_____
A Rh+	AB Rh+	O Rh+	_____
O Rh+	A Rh+	O rh–	_____
A Rh+	B Rh+	O rh–	_____
A MN	B N	B M	_____

13. In a mating $AB\ Rh+rh-\ MN \times AB\ Rh+rh-\ MN$, what is the probability that the first child will have the phenotype:

A rh– MN _____   AB Rh+ M _____

14. Chinese primrose flowers have a center ("eye") of a color different from the petals. Normally, the eye is of medium size and yellow. Variant eyes occur: large yellow ("Primrose Queen"), white ("Alexandria"), and blue ("Blue Moon"). Results of several crosses are:

HOMOZYGOUS PARENTS	F$_1$	F$_2$
Normal × Alexandria	Alexandria	3 Alexandria: 1 Normal
Alexandria × Primrose Queen	Alexandria	_____
Blue Moon × Normal	Normal	3 Normal: 1 Blue Moon
Primrose Queen × Blue Moon	Blue Moon	_____

   a. Arrange the phenotypes in order of relative dominance.

   b. Fill in the missing F$_2$ data in the above table.

Challenge Problems

15. The babies have been mixed up at the hospital. Please advise the staff on the correct placement of the babies based on the blood types as indicated below.

	BLOOD PHENOTYPES	
BABY		**COUPLE**
1. A MN Rh+	a.	A M Rh+ × B MN Rh+
2. O N Rh+	b.	A M Rh+ × A N Rh+
3. AB M rh–	c.	O N Rh– × AB N rh–
4. B N rh–	d.	A MN rh– × A MN Rh+

16. In a certain species of animal, the genetics of blood types and fur coloration were determined by the matings presented in part b below.

   a. Based on the matings in part b, fill in the blanks concerning blood types and fur coloration.

BLOOD TYPING

A1 = R blood group

A2 = S blood group

A3 = T blood group

A1A2 = type _____

A1A3 = type _____

A2A3 = type _____

FUR COLORATION

B = _____ (black or brown)

b = _____ (black or brown)

BB = _____

Bb = _____

bb = _____

b. Give all the genotypes for the following matings:

CROSS 1   black, type ST   ×   black, type S

↓

16 black, type S
8 black, type T
7 black, type ST

CROSS 2   brown, type S   ×   black, type T

↓

23 brown, type ST
25 black, type ST

CROSS 3   brown, type S   ×   brown, type T

↓

17 brown, type ST   8 black, type T
19 brown, type R   9 black, type S
15 brown, type S   7 black, type R
16 brown, type T   9 black, type ST

Give the phenotypes and expected phenotypic ratios for offspring of the following mating:

CROSS 4   *Bb A2A3*   ×   *Bb A1A3*

Team Problems

17. Using the guinea pig alleles mentioned in Problem 3, what are the most probable genotypes for the parents that produce the following offspring:

PHENOTYPES OF PARENTS	PHENOTYPES OF PROGENY			
	BLACK	SEPIA	CREAM	ALBINO
a. black × black	22	0	0	7
b. black × albino	10	9	0	0
c. cream × cream	0	0	34	11
d. sepia × cream	0	24	11	12
e. black × albino	13	0	12	0
f. black × cream	19	20	0	0
g. black × sepia	18	20	0	0
h. black × sepia	14	8	6	0
i. sepia × sepia	0	26	9	0
j. cream × albino	0	0	15	17

18. The blood type *phenotypes* for parents and offspring for five families are given. Indicate the *full genotype* for all parents.

   FAMILY 1   Parents' phenotypes: male AB Rh+ × female O Rh+ →
   Offspring: 3/8 A Rh+: 3/8 B Rh+: 1/8 A rh–: 1/8 B rh–

   FAMILY 2   Parents' phenotypes: male A Rh+ × female A rh– →
   Offspring: 3/4 A Rh+: 1/4 O Rh+

   FAMILY 3   Parents' phenotypes: male B Rh+ × female A rh– →
   Offspring: 1/4 AB Rh+: 1/4 A Rh+: 1/4 B Rh+: 1/4 O Rh+

   FAMILY 4   Parents' phenotypes: male B Rh+ × female A rh– →
   Offspring: 1/8 AB Rh+: 1/8 AB rh–: 1/8 A Rh+: 1/8 A rh–: 1/8 B Rh+:
   1/8 B rh–: 1/8 O Rh+: 1/8 O rh–:

   FAMILY 5   Parents' phenotypes: male B Rh+ × female A Rh+ →
   Offspring: 3/16 AB Rh+: 3/16 A Rh+: 3/16 B Rh+:
   3/16 O Rh+: 1/16 AB rh–: 1/16 A rh–: 1/16 B rh–: 1/16 O rh–

Solutions

1. a. 2/16 or 1/8. If t = the gene for thalassemia and T = its normal allele, the heterozygote Tt has a mild form of the anemia. The husband's genotype is $Tt\ I^A i$ (because his mother was blood type O) and the wife's, $Tt\ I^A I^B$. The probability of a child with no anemia (TT) is 1/4, and the probability of blood group A is 2/4 (either $I^A I^A$ or $I^A i$). $1/4 \times 2/4 = 2/16$.

   b. 2/16 or 1/8, because 2/4 (Tt) × 1/4 ($I^B i$) = 2/16

   c. 1/16, because 1/4 (tt) × 1/4 ($I^A I^B$) = 1/16

2.

	PARENTS	OFFSPRING
a.	yellow × yellow	2 yellow: 1 agouti with light belly (because $A^Y A^Y$ is lethal)
b.	yellow × agouti with light belly	2 yellow: 1 agouti with light belly: 1 black and tan
c.	black and tan × agouti with light belly	2 agouti with light belly: 2 black and tan
d.	yellow × yellow	2 yellow: 1 agouti (because $A^Y A^Y$ is lethal, and A^+ dominant over a^t)

3. a. all black
   b. all black
   c. all black
   d. all sepia
   e. 3 black: 1 albino
   f. 1 black: 1 sepia
   g. 2 black: 1 sepia: 1 cream

4. a. $BO \times BO$
   b. $BB \times AB$
   c. $BB \times AO$
   d. $BO \times AO$
   e. $BO \times AB$
   f. $BO \times OO$

5. Exclude all except b (A Rh+ MN).

6. a. 1
   b. 2
   c. 2
   Ten genotypes are possible in the population.

7. 30 (10 for the c locus × 3 for the d locus)

8. 3 dark bellied: 1 black

9. One parent is heterozygous for white bellied and dark bellied; the other parent is either the same or heterozygous for white bellied and plain black.

10. white bellied > dark bellied > black and tan > plain black

11. a. black (Cc) × black (Cc)
    b. sepia ($c^k c$) × cream ($c^d c$)
    c. black (Cc^d) × albino (cc)
    d. black (Cc^k) × cream ($c^d c^d$ or $c^d c$)
    e. black (Cc^d) × sepia ($c^k c^d$ or $c^k c$)

12.

MOTHER	CHILD	ACCUSED MALE	ELIMINATE MALE?
A Rh+	O Rh+	AB Rh+	yes
A Rh+	AB Rh+	O Rh+	yes
O Rh+	A Rh+	O rh–	yes
A Rh+	B Rh+	O rh–	yes
A MN	B N	B M	yes

13. A rh– MN   1 chance in 32
    AB Rh+ M   3 chances in 32

14. a.   Alexandria > Normal > Blue Moon > Primrose Queen

    b.

HOMOZYGOUS PARENTS	F$_1$	F$_2$
Normal × Alexandria	Alexandria	3 Alexandria: 1 Normal
Alexandria × Primrose Queen	Alexandria	**3 Alexandria: 1 Primrose Queen**
Blue Moon × Normal	Normal	3 Normal: 1 Blue Moon
Primrose Queen × Blue Moon	Blue Moon	**3 Blue Moon: 1 Primrose Queen**

15. Couple a gets baby 3.
    Couple b gets baby 1.
    Couple c gets baby 4.
    Couple d gets baby 2.

16. Notice that in cross 2, S × T → all ST; thus, the parents are homozygous (*A2A2* and *A3A3*) and S and T are codominant. Also, in cross 3, S × T → R, S, T, and ST in about equal numbers; thus, the parents are heterozygous (*A2A1* and *A3A1*), and R is recessive to both S and T. The offspring in cross 3 are 67 brown and 36 black (an approximate 2:1 ratio, indicative of recessive lethal for homozygous brown individuals).

    a.

BLOOD TYPING	FUR COLORATION
A1 = R blood group	*B* = **brown**
A2 = S blood group	
A3 = T blood group	*b* = **black**
A1A2 = type **S**	*BB* = **lethal**
A1A3 = type **T**	*Bb* = **brown**
A2A3 = type **ST**	*bb* = **black**

    b. **CROSS 1**   black, type ST   ×   black, type S

                    bb A2A3   ↓   *bb A1A2*
                16 black, type S   **bb A1A2 and bb A2A2**
                  8 black, type T   *bb A1A3*
                  7 black, type ST   *bb A2A3*

        CROSS 2   brown, type S   ×   black, type T

                    Bb A2A2   ↓   *bb A3A3*
                23 brown, type ST   *Bb A2A3*
                25 black, type ST   *bb A2A3*

        CROSS 3   brown, type S   ×   brown, type T
                    Bb A1A2   ↓   *Bb A1A3*

17 brown, type ST	*Bb A2A3*	8 black, type T	*bb A1A3*
19 brown, type R	*Bb A1A1*	9 black, type S	*bb A1A2*
15 brown, type S	*Bb A1A2*	7 black, type R	*bb A1A1*
16 brown, type T	*Bb A1A3*	9 black, type ST	*bb A2A3*

CROSS 4 Bb $A2A3$ $\times$ Bb $A1A3$

         brown, type ST $\downarrow$ brown, type T

2/16 brown, type S

4/16 brown, type T

2/16 brown, type ST

1/16 black, type S

2/16 black, type T

1/16 black, type ST

4/16 lethal

17. Remember that the allele that causes the phenotype of an offspring must appear in the genotype of at least one of the parents. Thus, in b, the sepia allele must occur in the black parent.

   a. $Cc^a \times Cc^a$

   b. $Cc^k \times c^a c^a$

   c. $c^d c^a \times c^d c^a$

   d. $c^k c^a \times c^d c^a$

   e. $Cc^d \times c^a c^a$

   f. $Cc^k \times c^d c^d$

   g. $Cc^k \times c^k c^k$

   h. $Cc^d \times c^k c^d$

   i. $c^k c^d \times c^k c^a$

   j. $c^d c^a \times c^a c^a$

18.

	FULL PARENTAL GENOTYPES				
PARENT	**FAMILY 1**	**FAMILY 2**	**FAMILY 3**	**FAMILY 4**	**FAMILY 5**
Male	AB	AO	BO	BO	BO
	$Rh + -$	$Rh + +$	$Rh + +$	$Rh + -$	$Rh + -$
Female	OO	AO	AO	AO	AO
	$Rh + -$	$Rh --$	$Rh --$	$Rh --$	$Rh + -$

Problems

1. Use chi-square analysis to answer the following:

   a. Suppose a coin is tossed 12 times and lands heads up 4 times and tails up 8 times. Is this a significant deviation from the expected 50:50 ratio?

   b. If the coin is tossed 120 times and lands heads up 40 times and tails up 80 times, is this a significant deviation from what is expected?

2. Consider the following cross in mice where the parents are taken from two homozygous strains:

   P: albino × agouti fur with black belly

   ↓

   F_1:   all agouti fur with white bellies

   ↓

   F_2:   46 albinos

   42 agouti fur with black bellies

   120 agouti fur with white bellies

   a. How many gene pairs are involved in this cross? Use chi-square analysis to support your hypothesis.

   b. Devise a hypothetical linear biochemical pathway to explain the results.

3. A rooster with a walnut comb is mated to 3 hens. Hen I has a pea comb and produces offspring in a ratio of 3 walnut: 3 pea: 1 rose: 1 single. Hen II is walnut combed and has only walnut-combed offspring. Hen III is walnut combed and has offspring in a ratio of 3 walnut: 1 rose. Determine the genotypes of the rooster and the 3 hens.

4. A mating between a homozygous black rat and a homozygous yellow rat will produce all gray offspring. When these gray F_1 rats were mated, their offspring consisted of 6 cream, 18 black, 57 gray, and 19 yellow animals.

   a. How many pairs of genes are interacting to produce color differences in these rats? Give generalized genotypes for all the colors mentioned.

   b. Use chi-square analysis to support your hypothesis.

5. Onion bulbs may be red, white, or yellow. These color differences are under the control of two pairs of genes that interact epistatically. The F_2 generation of a cross produced 40 yellow, 50 white, and 110 red onions.

   a. Give generalized genotypes for all three onion bulb colors.

   b. Give a possible mating sequence that could produce the F_2 generation mentioned above.

6. Within a certain species, plants may produce orange, red, or white flowers. In a series of matings, three different white-flowered plants were mated with three different orange-flowered plants. All plants were taken from the same population, and similar pigments are produced in similar ways in all the plants. The results of these matings are presented below.

	MATING 1	MATING 2	MATING 3
P:	white × orange	white × orange	white × orange
	↓	↓	↓
F_1:	all red	all orange	180 red
			160 orange
			350 white
	↓	↓	
F_2:	160 orange	290 orange	
	460 red	100 white	
	210 white		

   a. How many pairs of genes control flower color?
   b. Give generalized genotypes for the various flower colors.
   c. Give the genotypes for all plants in the three matings.

7. An X-linked gene in *Drosophila*, *v*, produces recessive vermilion eye color when homozygous in females or hemizygous in males. An autosomal gene, *bw*, produces recessive brown eye color. Both *vv bw bw* females and *v bw bw* males have white eyes. Determine the phenotypic expectations in the F_1 and F_2 generations when homozygous normal-eyed females are crossed to males that are white eyed due to the interaction of the vermilion and brown genes.

8. You were given a packet of 200 tomato seeds, all produced from matings involving the same parents. These seeds produced 12 plants with yellow fruits and hairless stems, 16 with yellow fruits and very hairy stems, 28 with yellow fruits and scattered short hairs on the stems, 32 with red fruit and very hairy stems, 38 with red fruit and hairless stems, and 74 with red fruits and scattered short hairs.

   a. Explain the inheritance of stem hairs and fruit color, and give the genotypes of the parents and the expected phenotypic ratios of their 200 offspring.
   b. Is chi-square analysis of the data consistent with your explanation?

9. In an insect species related to fruit flies, a cross between white-eyed males and pink-eyed females produced an F_1 generation where all had wild-type red eyes. Crosses between F_1 females and F_1 males produced the following F_2 individuals:

   red-eyed females   910
   white-eyed males   605
   red-eyed males   447
   pink-eyed females 290
   pink-eyed males   148

   Explain how red, pink, and white eye colors are inherited in this species.

10. If a man of genotype *Hh Sese $I^A I^A$* marries a woman of genotype *Hh Sese $I^B I^B$* and their children are tested to determine their ABO blood group status using A and B antibodies against blood samples and saliva samples, what are the phenotypic probabilities among the children?

11. In *Drosophila melanogaster*, a gene "ebony" produces a dark body color when homozygous, and an independently assorting gene "black" has a similar effect.
    a. Give the color of F_1 progeny from a cross between homozygous ebony and homozygous black.
    b. What phenotypes and proportions are expected in the progeny of an $F_1 \times F_1$ cross?
    c. What phenotypic ratios are expected among progeny of a cross between the F_1 and the ebony parental type?
    d. What phenotypic ratios are expected among progeny of a cross between the F_1 and the black parental type?

12. In sweet peas, genes *C* and *P* are necessary for purple flowers. In the absence of either or both (*ccpp*) genes, the flowers are white. Give the expected colors and proportions of offspring from the following crosses:
    a. *Ccpp × ccPp*
    b. *CcPp × Ccpp*
    c. *ccpp × CcPp*
    d. *CcPp × CcPp*

13. White Leghorn chickens are homozygous for a color gene (*CC*) and a dominant inhibitory gene (*II*) that prevents the action of *C*. White Wyandotte chickens are homozygous recessive at both loci (*iicc*). Give the F_2 phenotypes and proportions expected from a mating beginning with pure-bred Leghorn and Wyandotte parents (*IICC × iicc*).

14. Normal hearing in humans depends upon the presence of at least one dominant gene from each of two pairs, *D* and *E*. If you examined the collective progeny of many *DdEe × DdEe* marriages, what phenotypic ratio would be expected?

15. In sweet peas, the cross white flowers × white flowers produced an F_1 all with purple flowers. An F_2 of 350 white and 450 purple was then obtained.
    a. What is the phenotypic ratio of the F_2?
    Propose genotypes for:
    b. the purple F_2 plants
    c. the F_1 plants
    d. the two parental plants

16. In some plants, cyanidin, a red pigment, is synthesized enzymatically from a colorless precursor. Delphinidin, a purple pigment, may be made from cyanidin by the enzymatic addition of one —OH group to the cyanidin molecule. In one cross where these pigments were involved, purple × purple → 81 purple, 27 red, and 36 white.
    a. How many pairs of genes are involved?
    b. What are the genotypes of the purple parents?
    c. Give the genotypes of all the offspring.

17. In dingbats, toenail color may be red, white, or blue. A series of crosses was made to determine the mode of inheritance of toenail color in this species. Three separate crosses, the results of which are given below, were performed using parents that may or may not have been homozygous.

    a. Give the genotypes of all dingbats in the following matings. Use alphabet letters, starting with *A, a*.

red × blue	red × blue	red × white
___ ___	___ ___	___ ___
↓	↓	↓
all red	32 red ___	12 blue ___
___	14 white ___	45 red ___
	17 blue ___	35 white ___

    ↓

    61 white ___
    236 red ___
    19 blue ___   (ratio: _____)

    b. Determine a hypothetical metabolic pathway that could explain the inheritance of toenail color in dingbats. The pathway must be consistent with all the data.

18. In a certain species of plant, different varieties producing yellow, white, or orange flowers are known.

    a. In the matings presented below, fill in the *genotypes* (using symbols in alphabetical order) of the P, F$_1$, and F$_2$ plants:

P (ALL HOMOZYGOUS) →	F$_1$ →	F$_2$		
yellow × yellow	yellow	yellow		
___ ___	___	___		
yellow × yellow	orange			
___ ___	___			
orange × white	orange	365 orange, 41 white, 235 yellow		
___ ___	___	___	___	___

    b. In the matings below, give the expected F$_2$ generation phenotypic ratios:

P (ALL HOMOZYGOUS) →	F$_1$ →	YELLOW	ORANGE	WHITE
white × yellow	→ yellow × →	___	___	___
orange × white	→ orange × →	___	___	___
orange × yellow	→ orange × →	___	___	___

    c. Determine a hypothetical metabolic pathway that could explain the inheritance of flower color in this species. This pathway must be consistent with all the data.

Challenge Problems

19. A dingbat may have spots on its head only, on its belly only, on its back only, or on its tail only. The position of the spots is controlled by a series of multiple alleles, with H^a for head spots, H^b for belly spots, H^c for back spots, and H^d for tail spots. Results from the following crosses were used to establish the dominance relationships of the four alleles for spotting position:

HOMOZYGOUS PARENTS		OFFSPRING
tail spots × back spots	$\rightarrow$	all tail spots
head spots × belly spots	$\rightarrow$	all belly spots
head spots × tail spots	$\rightarrow$	all head spots

   a. Give the dominance relationships for the alleles of spotting.

   Lack of spotting is caused by an epistatic interaction involving a pair of genes (S, s) that are not alleles of the spotting genes. In fact, the H and S loci show independent assortment. Alleles at the H loci have their phenotypic effects only when the animals are SS or Ss. An ss dingbat cannot devolop spots even though spotting alleles are present at the H locus. Consider the series of matings below:

   P: $H^aH^a\ SS \times H^dH^d\ ss$ $H^bH^b\ SS \times H^cH^c\ ss$

                           $\downarrow$ $\downarrow$

   F$_1$: $H^aH^d\ Ss\ female$ × $H^bH^c\ Ss\ male$

   b. Give the phenotypes of the parents and the F$_1$ female and male.
   c. Give the spotting phenotypes and phenotypic ratios expected in the F$_2$ generation.

20. Consider the following family. Both parents have brown wavy hair. Among the children, hair colors are brown, blond, and red, and hair "shapes" are straight, wavy, and curly. It is thought that a single gene pair (C^1 for curly and C^2 for straight) control curliness, with wavy being the heterozygous condition. Hair color appears to be epistatically controlled by at least two independently assorting loci (alleles A and a and B and b), with red hair being epistatic to blond and brown, both of which are hypostatic.

   a. Give generalized genotypes for the three hair colors, and construct a linear metabolic pathway consistent with your hypothesis.
   b. For this family, determine the expected phenotypes and ratios of children for both hair "shape" and hair color.

21.

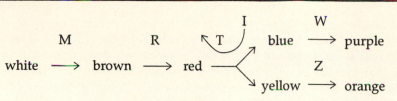

This is a hypothetical biochemical pathway for flower color in a plant species. Dominant genes *M, R, T, W,* and *Z* produce enzymes that catalyze the indicated reactions. Dominant gene *I* totally inhibits the effect of gene *T*. Recessive alleles (*i, m, r, t, w, z*) produce inactive gene products that do not catalyze reactions in the pathway. Example: plant *MMRRTTWWZZii* has purplish-orange flowers. In the following matings, give the phenotypes and ratios expected in the F_1 and F_2 generations:

a. *MMRRTTWWZZII × mmRRTTWWZZii*

b. *MMRRTTWWZZII × MMRRTTwwZZii*

c. *MMRRTTWWZZII × mmrrTTWWZZii*

Team Problems

22. In a certain species of plant, varieties producing white, purple, or red flowers are known. Study the results of the following matings, all beginning with homozygous parents.

	MATING 1	**MATING 2**	**MATING 3**
P:	red × red	red × red	purple × white
	↓	↓	↓
F_1:	all red	all purple	all purple
	↓	↓	↓
F_2:	all red		182 purple
			20 white
			118 red

a. How many gene pairs control color in this species? Support your answer by using chi-square calculations to analyze the F_2 ratio in mating 3.

b. Give generalized genotypes for each flower color.

c. Give the genotype for each plant in the three matings.

d. Construct a metabolic pathway for flower color consistent with the data.

e. For the three matings below, all beginning with homozygous parents, give the phenotypes and phenotypic ratios expected in the F_2 generation:

	MATING 4	**MATING 5**	**MATING 6**
P:	purple × red	white × red	purple × white
	↓	↓	↓
F_1:	all purple	all red	all purple

23. Coat color in mice is controlled by many genes, four of which we will consider. The genotype $C-$ permits pigment deposition in the fur, whereas cc causes albinism. The genotype $B-$ causes black pigmentation, whereas bb results in brown fur. The genotype $A-$ causes deposition of a narrow band of yellow pigment near the tips of individual hairs, a condition called "black agouti" in black animals and "cinnamon" in brown animals; genotype aa causes no yellow banding of hairs. The genotype dd causes a dilution of fur color (it has no effect on albinism); genotype $D-$ has no effect on coloration at all.

A cross occurred between mice of the genotypes given below:

$$CcBbaaDd \times CcbbAaDd$$

a. Give the phenotypes of these parental mice.

b. Give the generalized genotypes and expected phenotypic ratios of the following types of offspring produced from the above mating:

PHENOTYPE	GENERALIZED GENOTYPE	EXPECTED RATIO OF PHENOTYPE
black	_____	____
black agouti	_____	____
dilute black	_____	____
dilute black agouti	_____	____
brown	_____	____
dilute brown	_____	____
cinnamon	_____	____
dilute cinnamon	_____	____
albino	_____	____

24. In rats, two independently assorting pairs of genes, A, a and R, r, interact as follows: $A-R-$ = gray, $A-rr$ = yellow, $aaR-$ = black, and $aarr$ = cream. These hair colors are expressed only in the presence of a dominant third gene, D; its recessive allele, d, causes albinism.

a. Fill in the full genotypes, whenever possible, for all rats in the following table, using all three pairs of genes:

HOMOZYGOUS PARENTS $\rightarrow$	F_1 $\rightarrow$	NUMBERS OF F_2
gray × albino 1	all gray	174 gray, 65 black, 80 albino
gray × albino 2	all gray	48 gray, 16 albino
gray × albino 3	all gray	104 gray, 33 yellow, 44 albino
gray × albino 4	all gray	292 gray, 88 black, 171 albino, 87 yellow, 32 cream

b. If crosses between rats of the following genotypes produce a total of 1,000 offspring, indicate the expected numbers of each type of offspring.

$$AaRrDd \times AaRrDd \rightarrow \text{gray, yellow, black, cream, and albino}$$

Solutions

1. a.

CLASSES	OBSERVED	EXPECTED	$(o-e)$	$(o-e)^2$	$(o-e)^2$/EXP
heads	4	6	−2	4	$4/6 = 0.67$
tails	8	6	+2	4	$4/6 = \underline{0.67}$
	12	12	0		$\text{chi}^2 = 1.34$

Note: 1 degree of freedom, $0.30 > P > 0.20$.

The data are not significantly different from a 50:50 ratio.

b.

CLASSES	OBSERVED	EXPECTED	$(o-e)$	$(o-e)^2$	$(o-e)^2$/EXP
heads	40	60	−20	400	$400/60 = 6.67$
tails	80	60	+20	400	$400/60 = \underline{6.67}$
	120	120	0		$\text{chi}^2 = 13.34$

Note: 1 degree of freedom, $P < 0.001$.

The data are significantly different from a 50:50 ratio.

2. a. Two pairs, because a 9:3:4 ratio (which is a modified dihybrid 9:3:3:1 ratio) is obtained in the F_2 generation. Chi-square analysis is consistent with this hypothesis.

CLASSES	OBSERVED	EXPECTED	$(o-e)$	$(o-e)^2$	$(o-e)^2$/EXP
agouti/white	120	117	+3	9	0.08
agouti/black	42	39	+3	9	0.23
albino	48	52	−6	36	$\underline{0.69}$
	208	208	0		$\text{chi}^2 = 1.00$

Note: 2 degrees of freedom, $0.50 > P > 0.30$.

A hypothesis of one pair of genes acting to give a 1:2:1 ratio is rejected by chi-square analysis.

CLASSES	OBSERVED	EXPECTED	$(o-e)$	$(o-e)^2$	$(o-e)^2$/EXP
agouti/white	120	104	+16	256	2.46
agouti/black	42	52	−10	100	1.92
albino	48	52	+6	36	$\underline{0.69}$
	208	208	0		$\text{chi}^2 = 5.17$

Note: 1 degree of freedom, $P < 0.05$.

b. In a 9:3:4 ratio, the genotypes would be 9 *A–B–*: 3 *A–bb*: 4 *aaB–* and *aabb*. Because the presence of *aa* blocks the expression of *B–*, the A step precedes the B step in the pathway. The *aa––* phenotypes are albino, the *A–bb* phenotype is agouti with black belly, and the *A–B–* phenotype is agouti with white belly. Thus, the hypothetical linear biochemical pathway is:

enzyme A
from genotype *A–*

no pigment $\longrightarrow$ agouti with black belly

enzyme B
from genotype *B–*

$\longrightarrow$ agouti with white belly

3. In chickens, $R–P–$ = walnut, $R–pp$ = rose, $rrP–$ = pea, and $rrpp$ = single combs. Because hen I produces some single-combed offspring, the rooster must be heterozygous for both gene pairs ($RrPp$), and hen I must be $rrPp$, otherwise no $rrpp$ offspring would be possible:

$$\text{rooster (walnut)} \qquad \text{hen I (pea)}$$
$$RrPp \qquad \times \qquad rrPp$$
$$\downarrow$$

3/8 $RrP–$ (walnut)

3/8 $rrP–$ (pea)

1/8 $Rrpp$ (rose)

1/8 $rrpp$ (single)

Because hen II has only walnut-combed offspring, she must be homozygous ($RRPP$); any other genotype would produce non-walnut-combed offspring:

$$\text{rooster (walnut)} \qquad \text{hen II (walnut)}$$
$$RrPp \qquad \times \qquad RRPP$$
$$\downarrow$$

all $R–P–$ (walnut)

Hen III must be $RRPp$ because she produces some rose-combed offspring:

$$\text{rooster (walnut)} \qquad \text{hen III (walnut)}$$
$$RrPp \qquad \times \qquad RRPp$$
$$\downarrow$$

3/4 $R–P–$ (walnut)

1/4 $R–pp$ (rose)

4. a. The color differences can be explained on the basis of two pairs of genes because the ratio is 9:3:3:1:

$$\text{black} \qquad\qquad \text{yellow}$$
$$AAbb \qquad \times \qquad aaBB$$
$$\downarrow$$

gray ($AaBb$)

$$\downarrow$$

9/16 gray ($A–B–$)

3/16 black ($A–bb$)

3/16 yellow ($aaB–$)

1/16 cream ($aabb$)

b. Chi-square analysis:

CLASSES	OBSERVED	EXPECTED	$(o - e)$	$(o - e)^2$	$(o - e)^2/\text{EXP}$
gray	57	56.52	0.75	0.56	$0.56/56.25 = 0.01$
black	18	18.75	−0.75	0.56	$0.56/18.75 = 0.03$
yellow	19	18.75	0.25	0.06	$0.06/18.75 = 0.01$
cream	6	6.25	−0.25	0.06	$0.06/\ 6.25 = 0.01$
	100	100.00	0		$\text{chi}^2 = 0.06$

Note: 3 degrees of freedom, $P > 0.95$.

5. a. The numbers approximate a 9 red: 3 yellow: 4 white ratio. This is supported by chi-square analysis:

CLASSES	OBSERVED	EXPECTED	$(o - e)$	$(o - e)^2$	$(o - e)^2/\text{EXP}$
red	110	112.5	−2.5	6.25	0.06
yellow	40	37.5	2.5	6.25	0.17
white	50	50.0	0	0.00	0.00
	100	200.00	0		$\text{chi}^2 = 0.23$

Note: 2 degrees of freedom, $0.90 > P > 0.80$.

b. Two possible mating schemes are:

red	white	yellow	white
AABB ×	*aabb*	*AAbb* ×	*aaBB*

↓   ↓

red (*AaBb*)   red (*AaBb*)

↓   ↓

9/16 *A–B–* red   9/16 *A–B–* red
3/16 *A–bb* yellow   3/16 *A–bb* yellow
3/16 *aaB–* white   3/16 *aaB–* white
1/16 *aabb* white   1/16 *aabb* white

6. a. Two pairs of genes are involved because the F_2 ratio of mating 1 is approximately 9 red: 3 orange: 4 white, a modified dihybrid ratio. The ratio is determined by first adding the number of plants in the F_2: $460 + 210 + 160 = 830$, then dividing the total by 16: $830/16 =$ about 52, then dividing each number in the F_2 by 52: $460/52 =$ about 9, $160/52 =$ about 3, $210/52 =$ about 4. Hence, the 9:3:4 ratio. The other possibility, a 1:2:1 ratio based on one pair of gene differences, is rejected by chi-square analysis.

b. Relating the expected results of a dihybrid cross to the actual results in the F_2 of mating 1:

AaBb ×   *AaBb*   red ×   red (F_1 of mating 1)

↓   ↓

9/16 *A–B–*   9/16 red (*A–B–*)
3/16 *A–bb*   3/16 orange (*A–bb*)
3/16 *aaB–*   4/16 white (*aaB–* and *aabb*)
1/16 *aabb*

c.

	MATING 1	MATING 2	MATING 3
P:	white × orange	white × orange	white × orange
	$aaBB$　　$AAbb$	$aabb$　　$AAbb$	$aaBb$　　$Aabb$
	↓	↓	↓
F_1:	red	orange	1/4 red
	$(AaBb)$	$(Aabb)$	1/4 orange $(Aabb)$
			2/4 white $(aa\text{--}\text{--})$
	↓	↓	
F_2:	9/16 red $(A\text{--}B\text{--})$	3/4 orange $(A\text{--}bb)$	
	3/16 orange $(A\text{--}bb)$	1/4 white $(aabb)$	
	4/16 white $(aa\text{--}\text{--})$		

7.

P:　$v^+v^+\ bw^+bw^+$　×　$v\ Y\ bw\ bw$
　　(normal)　　　　　(white)

　　　　　　　↓

F_1:　$v^+v\ bw^+bw$　=　normal-eyed females
　　$v^+Y\ bw^+bw$　=　normal-eyed males

F_2:

	SPERM			
EGGS	$v^+\ bw^+$	$v^+\ bw$	$bw^+\ Y$	$bw\ Y$
$v^+\ bw^+$	normal females	normal females	normal males	normal males
$v^+\ bw$	normal females	brown females	normal males	brown males
$v\ bw^+$	normal females	normal females	vermilion males	vermilion males
$v\ bw$	normal females	brown females	vermilion males	white males

EYE COLOR	FEMALES	MALES	TOTAL
normal	6/8	3/8	9/16
brown	2/8	1/8	3/16
vermilion	0/8	3/8	3/16
white	0/8	1/8	1/16

8.　a.　Analyze each trait separately. There are 50 hairless, 102 short-haired, and 48 very hairy (1:2:1), and 144 red and 56 yellow (3:1). Let R represent red and r yellow (complete dominance). Let $h1h1$ = very hairy, $h1h2$ = scattered hairs, and $h2h2$ = hairless (incomplete dominance).

P:　$Rrh1h2$ (red scattered hairs)　×　$Rrh1h2$ (red scattered hairs)

　　　　　　　　　　↓

F_1:　　　　　　3/16 $R\text{--}h1h1$ = red very hairy
　　　　　　6/16 $R\text{--}h1h2$ = red scattered hairs
　　　　　　3/16 $R\text{--}h2h2$ = red hairless
　　　　　　1/16 $rrh1h1$ = yellow very hairy
　　　　　　2/16 $rrh1h2$ = yellow scattered hairs
　　　　　　1/16 $rrh2h2$ = yellow hairless

b. Chi-square analysis of the data:

CLASSES	OBSERVED	EXPECTED	(o – e)	(o – e)^2	(o – e)^2/EXP
yellow hairless	12	12.5	–0.5	0.25	0.02
yellow very hairy	16	12.5	3.5	12.25	0.98
yellow short hairs	28	25.0	3.0	9.00	0.36
red hairless	38	37.5	0.5	0.25	0.01
red very hairy	32	37.5	–5.5	30.25	0.81
red short hairs	74	75.0	–1.0	1.00	0.01
	200	200	0.0		chi^2 = 2.19

Note: 5 degrees of freedom, 0.90 > P > 0.80.

The data are consistent with the proposed explanation for the inheritance of these traits.

9. The ratios of the eye colors are:

EYE COLOR	NUMBERS			RATIOS*		
	MALES	FEMALES	TOTALS	MALES	FEMALES	TOTALS
red	447	910	1357	3/8	6/8	9/16
pink	148	290	438	1/8	2/8	3/16
white	605	0	605	4/8	0/8	4/16
totals	1200	1200	2400			

*Divide 2,400 by 16 = 150; then divide each number by 150.

The genetic explanation is that white is recessive due to an X-linked gene (*w*); *ww* females and *w* Y males are epistatic for white eyes, regardless of other genes present. Pink is recessive due to an autosomal gene (*p*); white is epistatic to pink, which is hypostatic.

10. *hhSese I^{A}I^A* × *HhSese I^{B}I^B*

		BLOOD	SALIVA
9/16	*H– Se– I^{A}I^B*	AB	AB
3/16	*H– sese I^{A}I^B*	AB	O
3/16	*hh Se– I^{A}I^B*	O	O
1/16	*hh sese I^{A}I^B*	O	O

11. a. ++*ee* × *bb*++ → +*b*+*e* = normal
    b. +*b*+*e* × +*b*+*e* → 9 + – + – (9 normal): 3 *bb*+–: 3 +–*ee*: 1 *bbee* (7 dark)
    c. 1 normal: 1 dark
    d. 1 normal: 1 dark

12. a. 1 purple: 3 white
    b. 3 purple: 5 white
    c. 1 purple: 3 white
    d. 9 purple: 7 white

13. 9 *I–C–* (white): 3 *I–cc* (white): 3 *iiC–* (colored): 1 *iicc* (white) = 13 white: 3 colored

14. 9 *D–E–* (normal): 3 *D–ee* (deaf): 3 *ddE–* (deaf): 1 *ddee* (deaf) = 9 normal: 7 deaf

15. a. 9:7
    b. *A–B–*
    c. *Aabb*
    d. *AAbb × aaBB*

16. a. 2, because the ratio is 9 purple: 3 red: 4 white
    b. *AaBb*
    c. If one assumes gene *A* is responsible for the enzyme converting the colorless precursor to cyanidin and *B* for the enzyme converting cyanidin to delphinidin:

       purple = *A–B–*; red = *A–bb*; white = *aaB–* and *aabb*.

17. a. | red | × | blue | | red | × | blue | | red | × | white |
       |-----|---|------|---|-----|---|------|---|-----|---|-------|
       | *AABB* | | *aabb* | | *AaBb* | | *aabb* | | *AaBb* | | *aaBb* |

↓		↓		↓
all red		32 red *AaBb, Aabb*		12 blue *aabb*
AaBb		14 white *aaBb*		45 red *AaB–, Aabb*
		17 blue *aabb*		35 white *aaB–*

       ↓

       61 white *aaB–*

       236 red *A–B–*

       19 blue *aabb* (ratio: 12 red: 3 white: 1 blue)

    b. Hypothetical metabolic pathway: gene X makes the enzyme to convert red to blue; gene B makes the enzyme to convert blue to white; gene A makes an inhibitor of enzyme X product, blocking conversion of red to blue.

       ⟋X A   B

       red ⟶ blue ⟶ white

18. a. | P (ALL HOMOZYGOUS) | → | F$_1$ | → | F$_2$ |
       |---|---|---|---|---|
       | yellow × yellow | | yellow | | yellow |
       | *AAbb*   *AAbb* or | | *AAbb* or | | *AAbb* or |
       | *aaBB*   *aaBB* | | *aaBB* | | *aaBB* |
       | yellow × yellow | | orange | | |
       | *AAbb*   *aaBB* | | *AaBb* | | |
       | orange × white | | orange | | 365 orange + 41 white + 235 yellow |
       | *AABB*   *aabb* | | *AaBb* | | *A–B–*   *aabb*   *A–bb, aaB–* |
       | | | | | (a 9 orange: 6 yellow: 1 white ratio) |

b.

P (ALL HOMOZYGOUS) → F₁ →	F₂		
	YELLOW	ORANGE	WHITE
white × yellow → yellow × →	3	0	1
orange × white → orange × →	6	9	1
orange × yellow → orange × →	1	3	0

19. a. $H^b > H^a > H^d > H^c$ (belly > head > tail > back spotting) in dominance

b. P: F_1 female's parents:
mother has head spots
father has no spots

F_1 male's parents:
mother has belly spots
father has no spots

F_1:   female: $H^a H^d$ Ss (head spots)
male: $H^b H^c$ Ss (belly spots)

c. F_2:

	SPERM			
EGGS	H^b S	H^b s	H^c S	H^c s
H^a S	belly	belly	head	head
H^a s	belly	no spots	head	no spots
H^d S	belly	belly	tail	tail
H^d s	belly	no spots	tail	no spots

no spots   = 4/16
head spots   = 3/16
back spots   = 0/16
belly spots   = 6/16
tail spots   = 3/16

20. a. Because two brown-haired parents produce brown, blond, and red-haired children, the possible genotypes may be:

P: $AaBb$ (brown) × $AaBb$ (brown)
↓

F_1:   brown   = A–B–
blond   = A–bb
red   = aB– and aabb

    A–   B–
red → blond → brown

b. P: brown, wavy   ×   brown, wavy
            $AaBb$ C^1C^2 $AaBb$ C^1C^2
            $Aa \times Aa$ $Bb \times Bb$ $C^1C^2 \times C^1C^2$
              ↓   ↓   ↓
            3/4 $A–$ 3/4 $B–$ 1/4 C^1C^1
                        ×   ×   2/4 C^1C^2
            1/4 aa 1/4 bb 1/4 C^2C^2

9/64 $A–B–C^1C^1$ = brown, curly

18/64 $A–B–C^1C^2$ = brown, wavy

9/64 $A–B–C^2C^2$ = brown, straight

3/64 $A–bbC^1C^1$ = blond, curly

6/64 $A–bbC^1C^2$ = blond, wavy

3/64 $A–bbC^2C^2$ = blond, straight

3/64 $aaB–C^1C^1$ = red, curly

1/64 $aabbC^1C^1$ = red, curly

6/64 $aaB–C^1C^2$ = red, wavy

2/64 $aabbC^1C^2$ = red, wavy

3/64 $aaB–C^2C^2$ = red, straight

1/64 $aabbC^2C^2$ = red, straight

21. a. *MMRRTTWWZZII* (brown) × *mmRRTTWWZZii* (white) → F_1 all brown → F_2 9 brown: 3 purplish-orange: 4 white

    b. *MMRRTTWWZZII* (brown) × *MMRRTTwwZZii* (bluish-orange) → F_1 all brown → F_2 12 brown: 3 purplish-orange: 1 bluish-orange

    c. *MMRRTTWWZZII* (brown) × *mmrrTTWWZZii* (white) → F_1 all brown → F_2 39 brown: 16 white: 1 purplish-orange

22. a. Two pairs of genes are involved because the F_2 ratio from mating 3 is approximately 9 purple: 6 red: 1 white, a modified dihybrid ratio. It is determined by dividing all the numbers by 20, the lowest number. Chi-square analysis supports this hypothesis:

CLASSES	OBSERVED	EXPECTED	$(o – e)$	$(o – e)^2$	$(o – e)^2$/EXP
purple	182	180	+2	4	16/180 = 0.09
white	20	20	0	0	0/20 = 0.00
red	<u>118</u>	<u>120</u>	<u>–2</u>	4	16/120 = <u>0.13</u>
	320	320	0		chi^2 = 0.22

Note: 2 degrees of freedom, P > 0.95.

    b. Relating the expected results from a dihybrid cross with the actual results from mating 3:

$AaBb$ × $AaBb$ purple × purple (F_1 of mating 3)
    ↓   ↓
9/16 $A–B–$ 9/16 purple ($A–B–$)
3/16 $A–bb$ 6/16 red ($A–bb$ and $aaB–$)
3/16 $aaB–$ 1/16 white ($aabb$)
1/16 $aabb$

c.

	MATING 1	**MATING 2**	**MATING 3**
P:	red × red *AAbb* *AAbb*	red × red *AAbb* *aaBB*	purple × white *AABB* *aabb*
	↓	↓	↓
F_1:	all red (*AAbb*)	all purple (*AaBb*)	all purple (*AaBb*)
	↓	↓	↓
F_2:	all red (*AAbb*)		9/16 purple (*A–B–*) 6/16 red (*A–bb* and *aaB–*) 1/16 white (*aabb*)

d. Hypothetical metabolic pathway:

```
              enzyme A
          from genotype A–
white   ─────────────────→   red-1
                                    mix  →  purple
white   ─────────────────→   red-2
              enzyme B
          from genotype B–
```

e.

	MATING 4	**MATING 5**	**MATING 6**
P:	purple × red *AABB* *AAbb*	white × red *aabb* *AAbb*	purple × white *AABB* *aabb*
	↓	↓	↓
F_1:	purple (*AABb*)	red (*Aabb*)	purple (*AaBb*)
	↓	↓	↓
F_2:	3/4 purple (*AAB–*) 1/4 red (*AAbb*)	3/4 red (*A–bb*) 1/4 white (*aabb*)	9/16 purple (*A–B–*) 6/16 red (*A–bb, aaB–*) 1/16 white (*aabb*)

23. a. *CcBbaaDd* = black; *CcbbAaDd* = cinnamon

b.

PHENOTYPE	GENERALIZED GENOTYPE	EXPECTED RATIO OF PHENOTYPE
black	*C–B–aaD–*	9/64
black agouti	*C–B–A–D–*	9/64
dilute black	*C–B–aadd*	3/64
dilute black agouti	*C–B–A–dd*	3/64
brown	*C–bbaaD–*	9/64
dilute brown	*C–bbaadd*	3/64
cinnamon	*C–bbA–D–*	9/64
dilute cinnamon	*C–bbA–dd*	3/64
albino	*cc–––*	16/64

24. a.

HOMOZYGOUS PARENTS	→	F$_1$	→	NUMBERS OF F$_2$		
gray *DDAARR*	× albino 1 *ddaaRR*	all gray *DdAaRR*		174 gray, *D–A–RR*	65 black, *D–aaRR*	80 albino *dd––RR*
gray *DDAARR*	× albino 2 *ddAARR*	all gray *DdAARR*		48 gray, *D–AARR*	16 albino *ddAARR*	
gray *DDAARR*	× albino 3 *ddAArr*	all gray *DdAARr*		104 gray, *D–AAR–*	33 yellow, *D–AArr*	44 albino *dd––aa*
gray *DDAARR*	× albino 4 *ddaarr*	all gray *DdAaRr*		292 gray, *D–A–R–* 87 yellow, *D–A–rr*	88 black, *D–aaR–* 32 cream *D–aarr*	171 albino, *dd––––*

b.   *AaRrDd* × *AaRrDd*

$$\downarrow$$

gray	(27/64)	× 1000 =	421.0
yellow	(9/64)	× 1000 =	140.6
black	(9/64)	× 1000 =	140.6
cream	(3/64)	× 1000 =	46.9
albino	(16/64)	× 1000 =	250.0
total:			1000

Sex-Linkage and Nondisjunction

Problems

1. Assume a gene for warped wings (W) in *Drosophila* that is dominant and X linked. Give the expected genotype and phenotypic ratios for the progeny of each of the following crosses:

   a. warped male × normal female

   b. warped heterozygous female × normal male

   c. warped heterozygous female × warped male

   d. warped homozygous female × normal male

2. In *Drosophila*, the mutation for ebony body color is recessive and autosomal (*e*), and the white-eyed mutation is recessive and X linked. Determine the phenotypes and phenotypic ratios of the F_1 and F_2 for the following crosses:

   a. white-eyed, ebony-bodied females × pure breeding wild-type males

   b. the reciprocal cross

3. In the pedigree below, assume that the trait shown by darkened symbols is caused by a sex-influenced autosomal gene dominant in males and recessive in females.

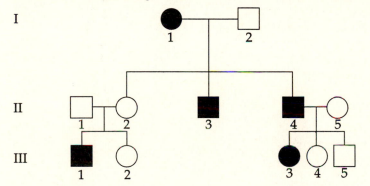

   a. What is the probability of producing an affected daughter from a mating between III-1 and III-4?

   b. What is the probability of producing an affected daughter from the mating of III-2 and III-5?

4. A family has 5 children, 3 girls and 2 boys. One of the boys died of Duchenne muscular dystrophy (X linked recessive) at age 15. The others are now adults and are concerned that their children will develop the disease. What would you tell them?

5. Pattern baldness is autosomal but sex influenced, being dominant in the male and recessive in the female. Red-green color blindness is an X-linked recessive trait. The following matings occur:

    P: bald   bald   bald   nonbald
           color-blind ×   non-color-blind   non-color-blind   ×   non-color-blind
           female   male   female   male

                          ↓   ↓

    F$_1$:   nonbald   bald
                         non-color-blind   ×   color-blind
                         female   male

    a. Give the genotypes of all these individuals.
    b. What types of offspring are expected in the F$_2$ generation?

6. In sheep, the autosomal gene for horned condition is dominant in males and recessive in females. If a hornless ram is mated to a horned ewe, what will be the chance that:

    a. an F$_2$ male sheep will be horned?
    b. an F$_2$ female will be horned?

7. An XXY *Drosophila* female, homozygous for forked bristles, is mated to a wild-type XY male. If, in this particular case, during female oogenesis there is 60% XX pairing and 40% XY pairing, what percentage of their progeny will consist of the following:

    a. inviable offspring
    b. males with forked bristles
    c. females with forked bristles
    d. normal-bristled males
    e. normal-bristled females

8. In a hypothetical mammal quite prone to nondisjunction, red blood cell (rbc) shape is controlled by X-linked alleles. The following mating produced the offspring indicated below. In each case, indicate the sex chromosome genotype (R = normal round rbc shape, r = oval shape). From the list of choices, pick probable explanations for the occurrence of each type of offspring.

    female with equal numbers of   male with all
    round and oval rbcs   ×   round rbcs

                                    ↓

    a.   female with all round rbcs
    b.   male with all oval rbcs
    c.   female with all oval rbcs
    d.   female with about 67% of the rbcs oval and 33% round
    e.   male with about 50% round and 50% oval rbcs

    Possible explanations in the parents:

    a) normal male meiosis, normal female meiosis
    b) normal male meiosis, anaphase I nondisjunction in female meiosis
    c) normal male meiosis, anaphase II nondisjunction in female meiosis
    d) normal female meiosis, anaphase I nondisjunction in male meiosis
    e) normal female meiosis, anaphase II nondisjunction in female meiosis

9. For the following humans, determine the sex chromosome constitutions and the number of chromosomes and Barr bodies in their somatic cells.

   a. normal female

   b. normal male

   c. Turner syndrome female

   d. Klinefelter male

   e. double-Y male

   f. triple-X female

10. Give the X/A ratio and sex designation for each of the following fruit flies (each A = 1 set of autosomes, each X = 1 X chromosome).

    a. AAXXXX

    b. AAAAAXX

    c. AAXXXXX

    d. AAAAAXXX

    e. AAAAXXXX

11. Abnormal eye shape in *Drosophila* may be caused by a variety of mutant genes, dominant or recessive, X linked or autosomal. One normal-eyed male from a true-breeding normal stock was crossed to two different abnormal-eyed females with the following results:

| | PROGENY OF FEMALE 1 | | PROGENY OF FEMALE 2 | |
	FEMALE	MALE	FEMALE	MALE
normal eyed	220	0	103	99
abnormal eyed	0	214	110	97

Explain the differences between these two sets of results, and give genotypes for the male and the two females that served as parents.

12. Hemophilia-A in humans is caused by an X-linked recessive gene.

    a. If a hemophiliac male mates with a homozygous nonhemophiliac female, what expected proportions and sexes among their offspring will be hemophiliacs?

    b. If a daughter produced by the mating in part a mates with a normal male, what expected proportions and sexes among their offspring will be hemophiliacs?

13. A woman of blood type A and normal color vision produced five children as follows:

    a. male, type A, color-blind

    b. male, type O, color-blind

    c. female, type A, color-blind

    d. female, type B, normal color vision

    e. female, type A, normal color vision

    Of the two men that mated with this woman at different times, the first is type AB and color-blind, and the second is type A and normal. For each child, which of the men is the most probable father?

14. A cross of white-eyed males × pink-eyed females → F_1 all with normal eye color. The $F_1 \times F_1$ mating produced F_2 flies as follows: 890 red-eyed females, 455 red-eyed males, 605 white-eyed males, 140 pink-eyed males, and 310 pink-eyed females. Define gene symbols, and give a genetic explanation consistent with the data.

15. In chickens, X-linked gene B (barred feathers) is completely dominant to b (nonbarred feathers). Autosomal gene R (rose comb) is completely dominant to its allele r (single comb).

    a. Give the F_1 phenotypes and ratio of the mating between barred homozygous rose comb female and nonbarred single-comb male.

    b. If $F_1 \times F_1$ occurs, give the sexes and the expected fraction of the F_2 that will be both barred and rose comb.

16. The dominant autosomal gene (B) for premature baldness in humans is sex limited. If a BB^+ male and a BB^+ female marry, what proportions of their (a) male and (b) female children are expected to become prematurely bald?

17. The gene for yellow body color in *Drosophila* is recessive and X linked. Its dominant allele produces normal gray body color. What phenotypic ratios (including sex) are expected from the following crosses?

    a. yellow male × yellow female

    b. yellow female × gray male

    c. homozygous gray female × yellow male

    d. heterozygous female × gray male

    e. heterozygous female × yellow male

18. At a hospital for the mentally retarded, a male attendant with Jacob (double-Y) syndrome mated with a female patient with triple-X syndrome. Figure out the sex and sex chromosome genotypes and syndrome names for all possible offspring. Assume that all nondisjunction possibilities occur at equal frequency, but that gametes without a sex chromosome are not produced.

19. A mating was performed involving XXY × XYY *Drosophila*, both of which are fully fertile. The parents as well as all offspring have a normal number of autosomes. Figure out the sex and sex chromosome constitution of all possible offspring. Also, give the expected sex ratio of the offspring. Assume that all nondisjunction possibilities occur at equal frequency, but that gametes without a sex chromosome are not produced.

20. A normal woman whose mother was color-blind and whose father had hemophilia (X-linked recessive) marries a color-blind man. They produce a son and a daughter. Ignoring the effect of crossing-over, what is the probability that:

    a. the son is color-blind but not hemophiliac?

    b. the daughter is phenotypically normal?

    c. the son is color-blind and has hemophilia?

    d. the son is normal for both traits?

    e. the daughter is color-blind only?

21. Primary generalized osteoarthritis (PGO) is caused by an autosomal gene that acts as a dominant in females and a recessive in males. If an unaffected man (whose father had the disease) marries an affected woman (who had an affected father and a normal mother), what is the probability that:

    a. their son will be affected?

    b. their son will be a carrier?

    c. their daughter will be affected?

    d. their daughter will be a carrier?

22. If a normal woman whose father has hemophilia (X linked recessive) marries a man with Marfan syndrome (a rare autosomal dominant), what is the chance that:

    a. any son will have hemophilia?

    b. any son will have Marfan?

    c. any son will have both traits?

    d. any daughter will have hemophilia?

    e. any daughter will have Marfan?

    f. any daughter will have both traits?

23. Assuming complete penetrance, could the trait indicated in the pedigree below be caused by (indicate yes or no):

    a. an autosomal gene recessive in males and females?

    b. an autosomal gene dominant in males and females?

    c. an autosomal gene dominant in males and recessive in females?

    d. an autosomal gene dominant in females and recessive in males?

    e. a Y-linked gene?

    f. an X-linked recessive gene?

    g. an X-linked dominant gene?

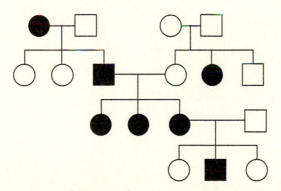

24. A color-blind (X-linked recessive) human with Klinefelter syndrome has parents both of whom have normal color vision.

    a. From which parent did he receive his extra X chromosome?

    b. Did nondisjunction occur during meiosis I or II in the parent who gave him the extra chromosome?

Challenge Problems

25. In cats, the following matings occurred, and the offspring indicated below were produced:

    P: female with patches of   male with all
                black and orange fur   ×   orange fur
                                    ↓
                        2 females with all orange fur
                        2 females with patches of black and orange fur
                        2 males with all black fur
                        2 males with all orange fur

    a. If fur color is controlled by X-linked alleles, give the exact genotypes of all these cats.

    b. What kinds of offspring would be expected (give genotypes and phenotypes) if the following meiotic abnormalities occurred involving the sex chromosomes?

        a) normal meiosis in the female parent but nondisjunction at anaphase I in the male parent

        b) normal meiosis in the male parent but nondisjunction at anaphase II in the female parent

        c) normal meiosis in the male parent but nondisjunction at anaphase I in the female parent

26. Pattern baldness is a sex-influenced trait, being autosomal dominant in males and autosomal recessive in females (A = baldness gene, a = nonbaldness gene). Color blindness and hemophilia both are X-linked recessive traits (E = normal color vision gene, e = color blindness gene; H = normal blood-clotting gene, h = hemophilia gene). Assume that no crossing-over occurs.

    a. Give complete genotypes, *including correct linkages to individual sex chromosomes*, for all individuals described in the P and F_1 generations.

    P: bald   nonbald   bald   bald
         non-color-blind   color-blind   color-blind   non-color-blind
         nonhemophilia   ×   hemophilia   nonhemophilia × hemophilia
         female   male   female   male
                       ↓   ↓
    F_1:   bald   nonbald
                  color-blind   non-color-blind
                  hemophilia   ×   nonhemophilia
                  male   female

    b. If the F_1 individuals above marry, give the probabilities of the various types of male and female offspring below, ignoring the effects of crossing-over:

        normal for all except hemophilia

        normal for all except color blindness and baldness

        normal for all except color blindness

        normal for all except baldness and hemophilia

        bald, color-blind, and with hempohilia

Team Problems

27. A common type of recessive red-green color blindness in humans is X linked.
    a. Can two color-blind parents produce
        a) a normal son?
        b) a normal daughter?
    b. Can two normal parents produce
        a) a color-blind son?
        b) a color-blind daughter?
    c. Can a normal daughter have
        a) a color-blind father?
        b) a color-blind mother?
    d. Can a color-blind daughter have
        a) a normal father?
        b) a normal mother?
    e. Can a normal son have
        a) a color-blind mother?
        b) a color-blind father?
    f. Can a color-blind son have
        a) a normal mother?
        b) a normal father?
    g. Can a color-blind brother and sister have
        a) another brother who is normal?
        b) another sister who is normal?
        c) parents who differ in their color vision phenotype (one color-blind, one not)?
    h. A normal woman whose father was color-blind marries a color-blind man. They produce a son and a daughter. What is the probability that
        a) the son is color-blind?
        b) the daughter is color-blind?

28. Assume allele *G* governs the dominant phenotype "striped," and allele *g* governs the recessive phenotype "unstriped." For each of the cases described below, give the phenotype of the male and female progeny.
    a. The locus is X linked in an animal with the XX-XO mechanism of sex determination, and unstriped females are mated with striped males.
    b. The locus is Z linked in an animal with the ZZ-ZW mechanism of sex determination, and striped females are mated with unstriped males.
    c. The locus is incompletely sex linked in an animal with the XX-XY mechanism of sex determination, and unstriped females are mated with striped males whose fathers were unstriped.
    d. The locus is incompletely sex linked in an animal with the ZZ-ZW mechanism of sex determination, and unstriped males are mated with striped females whose fathers were unstriped.
    e. The locus is autosomal in an animal with the ZZ-ZO mechanism of sex determination, but expression of the striped phenotype is sex limited to females, and unstriped females are mated to males with the *GG* genotype.

29. A young female circus acrobat who hangs by her hair as part of her act is wondering whether she should change her profession, if necessary, before it becomes too late. Her problem is this: her mother is bald, but her father has a normal head of hair. However, her older brother is rapidly losing his hair and will soon be bald. Let B represent the baldness gene and b represent the non-baldness gene. In humans, baldness is a sex-influenced trait. For each of the following questions, circle the correct answer:

  a. Which of these represents the parents of the young female circus acrobat?
- a) $XXBb \times XYBb$
- b) $XXbb \times XYBB$
- c) $X^bX^b \times X^BY$
- d) $XXBB \times XYbb$
- e) $X^BX^B \times X^bY$
- f) $XXBb \times XYbb$

  b. The genotype of the young female acrobat's older brother is:
- a) X^BY
- b) X^bY
- c) $XYBB$
- d) $XYBb$
- e) $XYbb$

  c. The young female acrobat's genotype is:
- a) X^BX^B
- b) $XBXb$
- c) $XbXb$
- d) $XXBB$
- e) $XXBb$
- f) $XXbb$

  d. On the basis of the data, which of the following suggestions to the young female acrobat is most justified?
- a) You need not change your profession—you will not become bald.
- b) The chances are 1 in 4 that you will become bald.
- c) The chances are 3 in 4 that you will become bald.
- d) There is a 50% chance that you will become bald.
- e) Change your profession—you will certainly become bald.

30. Sometimes nondisjunction will occur in a chromosomally normal parent and result in an aneuploid offspring. Often, the phenotype of the aberrant offspring will allow geneticists to determine in which parent and during which division of meiosis the nondisjunction occurred. For each of the four families described below, answer the questions. (Assume that all parents have a normal diploid chromosome complement.)

FAMILY 1   A man with the X-linked dominant condition "brown teeth enamel" marries a normal woman. They produce a son with brown tooth enamel.

FAMILY 2   A man makes 0 units of an enzyme due to homozygous recessive genes on chromosome 21. His wife makes 100 units of the enzyme due to a heterozygous pair of genes on chromosome 21. Their son has three copies of chromosome 21 and makes 200 units of the enzyme.

FAMILY 3 The parents in family 1 produce another son, this time with two Y chromosomes and normal tooth enamel.

FAMILY 4 Parents with "heterochromatic centromeres" for chromosome 13 produce a male child with three different versions of chromosome 13 as diagrammed below:

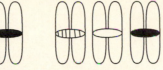

male parent   *female parent*   *child*

a. In which parent did the nondisjunction occur?

b. Did nondisjunction occur at meiosis I or II?

c. What is the number of sex chromosomes in the child's somatic cells?

d. What is the name of the syndrome in the child?

e. Will the child be able to reproduce?

Solutions

1.

	MALE	×	**FEMALE**	→	**MALE**	**FEMALE**
a.	W Y	×	++		all + Y normal	all W+ warped
b.	+ Y	×	W+		1/2 + Y normal 1/2 W Y warped	1/2 ++ normal 1/2 W+ warped
c.	W Y	×	W+		1/2 + Y normal 1/2 W Y warped	1/2 WW warped 1/2 W+ warped
d.	+ Y	×	WW		all W Y warped	all W+ warped

2. a. F_1 females: normal (wild type); F_1 males: white eyed

   F_2 males and females: 3/8 wild type: 1/8 ebony: 3/8 white: 1/8 white ebony

   P   *ww ee*   ×   *+ Y ++*

   F_1 *+w +e*   *w Y +e*

   F_2

	FEMALES		**MALES**	
	1/2 +w	↗ 3/4 +− ↘ 1/4 *ee*	1/2 + Y	↗ 3/4 +− ↘ 1/4 *ee*
	1/2 *ww*	↗ 3/4 +− ↘ 1/4 *ee*	1/2 *w* Y	↗ 3/4 +− ↘ 1/4 *ee*

b. F_1 females and males: wild type

F_2 females: 6/8 wild type: 2/8 ebony

F_2 males: 3/8 wild type: 1/8 ebony body: 3/8 white eyes: 1/8 white eyes and ebony body

P   ++ ++   × w Y ee

             ↓

F_1 $+w$ $+e$ $+$ Y $+e$

             ↓

F_2

FEMALES		MALES	
	↗ 3/4 +−		↗ 3/4 +−
1/2 ++		1/2 + Y	
	↘ 1/4 ee		↘ 1/4 ee
↗ ↘		↗ ↘	
	↗ 3/4 +−		↗ 3/4 +−
1/2 +w		1/2 w Y	
	↘ 1/4 ee		↘ 1/4 ee

3. Let T = trait and T' = normal.

MALES		FEMALES
trait	TT	trait
trait	TT'	normal
normal	$T'T'$	normal

a. Affected females must be TT. In this case, the chance of a TT female is the chance that III-4 is heterozygous (2/3 because both II-4 and 5 are TT') times the probability of TT from a mating of heterozygotes (1/4) times the chance that the child will be a girl (1/2). So, 2/3 × 1/4 × 1/2 = 1/12.

b. Because III-5 is $T'T'$ and a female must be TT to express the trait, this mating cannot produce an affected daughter.

4. Let D = normal gene and d = muscular dystrophy gene.

        Dd × D Y

             ↓

	SPERM	
EGGS	**D**	**Y**
D	DD normal	DY normal
d	Dd normal carrier	dY affected

There is no chance that any children of the boy will develop the disease as long as he marries a DD girl. Each girl has a 0.5 probability of being heterozygous and, therefore, transmitting the disease to half her sons.

5. a.   *rrBB* × *R* Y *Bb*   *RrBB* × *R* Y *bb*

             ↓   ↓

         RrBb   ×   *r* Y *Bb*

             ↓

		SPERM		
EGGS	*r B*	*r b*	Y *B*	Y *b*
R B	non-cb* bald females	non-cb nonbald females	non-cb bald males	non-cb bald males
R b	non-cb nonbald females	non-cb nonbald females	non-cb bald males	non-cb nonbald males
r B	cb bald females	cb nonbald females	cb bald males	cb bald males
r b	cb nonbald females	cb nonbald females	cb bald males	cb nonbald males

*cb = color-blind

6. P: *hh* male × *HH* female

               ↓

    F_1: *Hh* (horned) males, *Hh* (hornless) females

    a. 3/4, because horned is dominant in males. He may be heterozygous or homozygous (1/4 *HH* + 2/4 *Hh*).

    b. 1/4, because horned is recessive in females. She must be homozygous (1/4 *hh*).

7. XX pairing and segregation produce all forked gametes, of which 1/2 carry a Y chromosome: 0.60 (60%) X-X, Y segregation during oogenesis: 0.30 *f* and 0.30 *f* Y

    XY pairing and segregation produce forked and Y gametes. The unpaired X with forked assorts independently with the other X half the time and with the Y half the time: 0.40 (40%) X-Y, X segregation during oogenesis: 0.10 *f f*, 0.10 Y, 0.10 *f*, and 0.10 *f* Y

    The combination of these gametes with normal sperm is:

	EGGS			
SPERM	0.4 *f*	0.4 *f* Y	0.1 *f f*	0.1 Y
0.5 *f*+	0.20 *f*+ *f*	0.20 *f*+ *f* Y	0.05 *f*+ *f f* die	0.05 *f*+
0.5 Y	0.20 *f* Y	0.20 *f* Y Y	0.05 *f f*	0.05 Y Y die

    a. 10%

    b. 40%

    c. 5%

    d. 5%

    e. 40%

8. *Rr* × *R* Y

      ↓

    a. *RR*   choice a

    b. *r* Y   choice a

    c. *r*   choice d or e

    d. *Rrr*   choice c

    e. *Rr* Y   choice b or d

9.

	SEX CHROMOSOMES	CHROMOSOME NUMBER	BARR BODY NUMBER
a.	XX	46	1
b.	XY	46	0
c.	XO	45	0
d.	XXY	47	1
e.	XYY	47	0
f.	XXX	47	2

10.

	X/A RATIO	SEX
a.	2.0	metafemale
b.	0.4	metamale
c.	2.5	metafemale
d.	0.6	intersex
e.	1.0	female

11. Female 1 is homozygous for the X-linked recessive abnormal eye gene. Female 2 is heterozygous for the autosomal dominant abnormal eye gene.

12. a. None will be a hemophiliac.
    b. All daughters will be normal; half the sons will be normal, and half will be hemophiliacs.

13. a. either male
    b. male 2
    c. male 1
    d. male 1
    e. either male

14. white = X-linked recessive, pink = autosomal recessive; white shows recessive epistasis to pink:

        W–P– and W Y P– = red

        W–pp and W Y pp = pink

        $ww$$P$– and w Y P– = white

        $wwpp$ and w Y pp = white

F_1: $WwPp \times W$ Y Pp

      ↓

F_2:   6 red females: 2 pink females: 3 red males: 1 pink male: 4 white males

15. a. 1 barred rose male: 1 nonbarred rose female
    b. 6/16, equally divided between male and female

16. a. 3/4 of the sons
    b. none of the daughters

17. a. all yellow offspring
    b. gray females and yellow males
    c. all gray offspring
    d. all gray females, 50% yellow males, 50% gray males
    e. 50% yellow females, 50% gray females, 50% yellow males, 50% gray males

18.

	SPERM			
EGGS	XY	Y	X	YY
X	XXY Klinefelter male	XY normal male	XX normal female	XYY Jacob male
XX	XXXY Klinefelter male	XXY Klinefelter male	XXX triple-X female	XXYY Klinefelter male

19.

	SPERM			
EGGS	XY	Y	X	YY
X	XXY female	XY male	XX female	XYY male
XY	XXYY female	XYY male	XXY female	XYYY male
XX	XXXY female	XXY female	XXX female	XXYY female
Y	XYY male	YY lethal	XY male	YYY lethal

Note: sex ratio: 8 females: 6 males (plus 2 lethals).

20. a. 50% chance
    b. 50% chance
    c. no chance
    d. no chance
    e. 50% chance

21. a. 25% chance
    b. 50% chance
    c. 75% chance
    d. no chance

22. a. 50% chance
    b. 50% chance
    c. 25% chance
    d. no chance
    e. 50% chance
    f. no chance

23. a. yes
    b. no
    c. no
    d. ye
    e. no
    f. no
    g. no

24. a. $X^E Y \times X^E X^e \rightarrow X^e X^e Y$; from the female parent
    b. meiosis II

25. Due to random X chromosome inactivation,

    a. $BO \times O\,Y$

    $\downarrow$

OO	orange female
BO	black and orange female
B Y	black male
O Y	orange male

    b. a)

BO Y	black and orange male
OO Y	orange male
B	black female
O	orange female

    b)

BBO	black and orange female
OOO	orange female
BB Y	black male
OO Y	orange male
O	orange female
Y	lethal

    c)

BOO	black and orange female
BO Y	black and orange male
O	orange female
Y	lethal

26. a. P:

bald		nonbald		bald		bald	
non-color-blind		color-blind		color-blind		non-color-blind	
nonhemophilia	×	hemophilia		nonhemophilia	×	hemophilia	
female		male		female		male	
$AAX^{EH}X^{eh}$		$aaX^{eh}\,Y$		$AAX^{eH}X^{e-}$		$AaX^{Eh}\,Y$	

    $\downarrow$ $\downarrow$

    F₁:

bald		nonbald
color-blind		non-color-blind
hemophilia	×	nonhemophilia
male		female
$AaX^{eh}\,Y$		$AaX^{eH}X^{Eh}$

    $\downarrow$

b.

PHENOTYPE	PROBABILITIES AMONG		
	DAUGHTERS ONLY	SONS ONLY	ALL CHILDREN
normal for all except hemophilia	3/8	1/8	4/16
normal for all except color blindness and baldness	1/8	3/8	4/16
normal for all except color blindness	3/8	1/8	4/16
normal for all except baldness and hemophilia	1/8	3/8	4/16
bald, color-blind, and with hemophilia	0	0	0

27. a. a) no, b) no
    b. a) yes, the mother could be a carrier
       b) no, because the father is normal
    c. a) yes, b) yes
    d. a) no, b) yes, the mother could be a carrier
    e. a) no, b) yes
    f. a) yes, b) yes
    g. a) yes, b) yes, c) yes; in all three cases, the mother could be a carrier
    h. $Rr \times r\,Y$

EGGS	SPERM	
	r	Y
R	Rr non-color-blind	R Y non-color-blind
r	rr color-blind	r Y color-blind

a) 1/2, b) 1/2

28.

PARENTS		OFFSPRING	
MALES	FEMALES	MALES	FEMALES
a. G O	gg	unstriped	striped
b. gg	G W	striped	unstriped
c. G (Y g)	gg	unstriped	striped
d. gg	G (Z g)	unstriped	striped
e. GG	g O	unstriped	striped

29. a. d) The bald mother must be homozygous for the balding gene, and the nonbald father must be homozygous for the nonbalding gene.

   b. d) All sons must be heterozygous.

   c. e) All daughters must be heterozygous.

   d. a) Heterozygous females will not become bald because the baldness gene acts as a recessive in females.

30.

	FAMILY 1	FAMILY 2	FAMILY 3	FAMILY 4
a.	male	female	male	female
b.	meiosis I	meiosis II	meiosis II	meiosis I
c.	XXY = 3	XY = 2	XYY = 3	XY = 2
d.	Klinefelter syndrome	trisomy-21 (Down syndrome)	Jacob (double-Y) syndrome	Patau syndrome
e.	no	no	yes	no

7
Eukaryotic Gene Mapping

Problems

1. In corn, colored aleurone (R) is dominant to colorless (r), and green plant color (G) is dominant to yellow (g). Two plants, each heterozygous for both characteristics, are testcrossed to homozygous recessives, and their progeny are combined to produce the following totals:

colored green	100
colored yellow	97
colorless green	103
colorless yellow	100

    a. Use chi-square analysis to test these data for independent assortment of the two characteristics.
    b. When the progeny of each of the two heterozygous plants are scored separately, the following results are obtained:

PHENOTYPES	PLANT 1	PLANT 2
colored green	88	12
colored yellow	12	85
colorless green	8	95
colorless yellow	92	8

    Use chi-square analysis to test each data set for independent assortment.
    c. Explain the results of the three chi-square analyses.

2. A student has two dominant traits dependent on single genes, cataract (an eye abnormality), which he inherited from his mother, and polydactyly (extra fingers and/or toes), which he inherited from his father. If the loci for these two traits are very closely linked, would the student's child be more likely to have:
    a. either cataract or polydactyly?
    b. both cataract and polydactyly?
    c. neither trait?
    Explain your answers.

3. Mutant genes at two X-linked loci in *Drosophila* produce the recessive traits crossveinless wings (*cv*) and singed bristles (*sn*). The gametes produced by a heterozygous female are:

GENOTYPES	NUMBERS
cv sn	2,435
$+^{cv} +^{sn}$	2,460
cv $+^{sn}$	56
$+^{cv}$ *sn*	49

   a. Is the female in the coupling or repulsion phase of linkage?
   b. What is the map distance between the two loci?
   c. If the *cv* locus is 10.9 map units from the left end of the chromosome and the *sn* locus is closer to the centromere, what is the precise location of the *sn* locus?

4. In *Drosophila melanogaster*, the recessive gene *sr* (striped body) and *e* (ebony body color) are located at map distances 62 and 70 map units, respectively, from the left end of the third chromosome. (Some textbooks refer to the map unit [mu] as a centi-Morgan [cM]. These are equivalent terms equal to 1% crossing-over.) A striped female (*sr* $+^e$/*sr* $+^e$) was mated with a male with ebony body ($+^{sr}$ *e*/$+^{sr}$ *e*).

   a. What kinds of gametes will be produced by the F_1 female and in what proportion?
   b. If these F_1 females are mated with normal males heterozygous in the repulsion stage for *sr* and *e*, what phenotypes would be expected and in what proportion?

5. Below are listed the phenotypes of one of the most frequent progeny and one of the least frequent progeny from each of four testcrosses. For each, designate the recessive gene in the center of the sequence.

	TESTCROSS	MOST FREQUENT PHENOTYPE	LEAST FREQUENT PHENOTYPE
a.	I	r L E	R L E
b.	II	R s Q	r S Q
c.	III	F m a	M a F
d.	IV	p X n	x P n

6. In *Drosophila*, three loci on the third chromosome are defined by the mutations *p* (pink eye color), *Pr* (prickly bristles), and *Cu* (curled wings). Females heterozygous for the three mutations are testcrossed and produce the following progeny:

PHENOTYPES OF TESTCROSS PROGENY			NUMBER OF
EYES	WINGS	BRISTLES	INDIVIDUALS
normal	normal	normal	2
pink	normal	normal	195
normal	normal	prickly	17
normal	curled	normal	796
pink	curled	normal	18
pink	normal	prickly	784
normal	curled	prickly	185
pink	curled	prickly	3

a. What was the phenotype of the males used for the testcross?
b. Which gene is in the middle of the sequence?
c. Map the three gene loci.
d. Determine the coefficient of coincidence.
e. Determine the coefficient of interference.

7. In a testcross, females heterozygous for genes r, s, and q (RrSsQq) are crossed to homozygous recessive males (rrssqq). The linkage relationships, if any, of these three loci are not known, nor is the coupling/repulsion relationship of the linked genes in the heterozygous female. The data below represent the offspring from this cross:

PHENOTYPES	NUMBERS
r s Q	170
r s q	80
R s q	170
R s Q	80
r S Q	170
r S q	80
R S q	170
R S Q	80
	1,000

a. Give the coupling/repulsion arrangement of these genes in the heterozygous female.
b. Construct a genetic map for these three loci.

8. In *Drosophila*, the map of three recessive X-linked mutations, v (vermilion eyes), s (sable body color), and sd (scalloped wings), is shown below:

v	s	sd
33.0	43.0	51.5

Vermilion, sable females are mated to scalloped males, and the F_1 heterozygous females are test-crossed. Assuming no interference in this region of the chromosome, if 10,000 testcross progeny are produced, how many should have the following phenotypes?
a. vermilion eyes, otherwise normal
b. scalloped wings, otherwise normal
c. wild type
d. sable body color, otherwise normal
e. Recalculate your answers for the above assuming a coefficient of interference of 0.7.

9. Three gene pairs (Aa, Bb, Cc) influence height in a particular plant such that each active gene (in capital letters) adds 2 cm, to a "base height" of 2 cm, and the three loci are linked as shown below:

A	9 mu	B	10 mu	C

If the F_1 progeny of a cross between two homozygous stocks ABC/ABC (14 cm) × abc/abc (2 cm) is crossed to the 2 cm parent, what will be the distribution of heights expected among the offspring? Assume no interference.

10. Four man-mouse hybrid subclones are shown below along with the human chromosomes remaining in each. Each subclone was tested for the activities of five human enzymes (a–e). Presence of enzyme activity is represented by "+" and absence by "−". Determine in each case the human chromosomes responsible for enzymes.

SUBCLONE	HUMAN CHROMOSOMES PRESENT	A	B	C	D	E
1	5, 17, 21	+	−	−	−	+
2	2, 5	−	+	−	−	−
3	2, 20, 17	+	+	−	+	+
4	4, 5, 20	−	v	+	+	−

11. Both mouse and human cells possess a particular enzyme, but the enzyme differs in electrophoretic migration between the two species. Somatic cell hybridization is performed, and the extracts of the subclones are subjected to electrophoresis and stained for the enzyme. The results are shown below. Which human chromosome contains the structural locus for the enzyme?

SUBCLONE	HUMAN CHROMOSOMES PRESENT	GEL BANDING PATTERNS
		origin —band movement— →
1	2, 4, 8, 21	
2	2, 6, 8, 9, 18	
3	2, 6, 17, 21	

12. If the recombination frequency between the linked autosomal genes A and B in *Drosophila* is 15%, what phenotypic frequencies are expected from a cross between AB/ab female × AB/ab male?

13. In mice, a strain of homozygous recessive frizzy-haired individuals is mated to a strain of homozygous recessive shakers (*ffSS* × *FFss*). The F_1 of this cross are intermated to produce an F_2 generation. If these gene loci are linked with a recombination frequency of 15% between them, what are the expected phenotypes and frequencies in the F_2 generation?

14. If genes A and B in *Drosophila* are X linked with 15% recombination between their loci, what phenotypic frequencies are expected in each sex among the progeny of each of the following crosses?
    a. *AB/ab* female × *ab* Y male
    b. *Ab/aB* female × *ab* Y male
    c. *AB/ab* female × *AB* Y male
    d. *AB/ab* female × *Ab* Y male
    e. *Ab/aB* female × *Ab* Y male

15. A strain homozygous for three recessive genes (*a*, *b*, and *c*), not necessarily linked in the same chromosome, is mated to a strain homozygous for its dominant normal alleles. F_1 females are then testcrossed with homozygous recessive males, and the following F_2 results are obtained:

PHENOTYPE	NUMBER
a b c	42
A B C	48
a B C	46
A b c	44
a b C	6
A B c	4
a B c	5
A b C	5

a. Which genes are linked?

b. How many map units apart are the linked loci?

16. In *Drosophila*, a cross between triply abnormal females with X-linked traits yellow bodies (recessive, *y*), echinus bristles (recessive, *ec*), and white eyes (recessive, *w*), and normal hemizygous males produces phenotypically normal F$_1$ females. These F$_1$ females are then testcrossed, and 1,000 offspring occur as follows:

	PHENOTYPES		
BODY	**BRISTLES**	**EYES**	**NUMBER**
normal	normal	normal	475
yellow	echinus	white	469
yellow	normal	normal	8
normal	echinus	white	7
yellow	normal	white	18
normal	echinus	normal	23
normal	normal	white	0
yellow	echinus	normal	0

Determine the order of the three loci in the X chromosome, and make a genetic map.

17. In a certain diploid plant, three loci (*Bb*, *Dd*, and *Gg*) are linked as follows:

B	20 mu	D	30 mu	G

There is only one plant to work with (the "parental plant"). It has the genotypic constitution *Bdg/bDG*.

a. If the plant is self-fertilized, what proportion of the progeny will be of genotype *bbddgg*?

b. If the "parental plant" is crossed with the *bbddgg* plant and 1,000 offspring are produced, list the expected numbers of each phenotypic type of offspring given below:

BDG   BdG   Bdg   bdG   BDg   bDG   bDg   bdg

18. In minks, a homozygous black-colored (*BB*), wavy-haired (*HH*), short-haired (*SS*) female was mated with a homozygous white (*bb*), straight (*hh*), long-haired (*ss*) male. The black, short, wavy-haired (*BbHhSs*) F$_1$ females produced the following offspring when testcrossed with white, straight, short-haired (*bbhhss*) males:

PHENOTYPES	NUMBERS (TOTAL = 800)
black, straight, long	42
white, wavy, short	38
black, straight, short	7
black, wavy, short	348
white, wavy, long	9
white, straight, long	356

a. Which gene is in the middle of the sequence?

b. Construct the genetic map.

19. A cross was made between an *aaddGG* female and an *AADDgg* male. The resulting F_1 females were testcrossed, and 1,000 offspring (the F_2 generation) were produced. The F_2 phenotypes and numbers are:

PHENOTYPE	NUMBERS
A D g	197
a D G	201
a d g	45
a D g	51
A d G	49
A D G	55
A d g	199
a d G	203

a. Are genes *D* and *G* linked? Explain.

b. Are genes *A* and *D* linked?

c. Are genes *A* and *G* linked?

d. How many linkage groups are represented?

e. Draw the complete genetic map, including all map distances.

A series of two-point mapping crosses indicated the following map distances involving the *A, D,* and *G* genes analyzed above and the additional genes *B, C, E,* and *F*:

C–A = 7 mu

A–E = 5 mu

B–F = 20 mu

D–F = 10 mu

C–E = 12 mu

C–G = 27 mu

D–B = 30 mu

f. Construct a map for genes *A, B, C, D, E, F,* and *G*, showing the correct gene sequences and map unit distances between loci.

g. How many linkage groups are represented in this expanded map?

Challenge Problems

20. The symbols *b*, *h*, and *m* represent three recessive genes linked on one of the autosomes of *Drosophila*. Females heterozygous for the three loci were crossed with homozygous recessive males. The following progeny were produced from this cross:

PHENOTYPES	NUMBERS
b m⁺ h⁺	235
b m h	130
b⁺ m h	245
b⁺ m⁺ h⁺	140
b m h⁺	35
b m⁺ h	90
b⁺ m⁺ h	45
b⁺ m h⁺	80

a. What is the coupling/repulsion arrangement of these genes in the heterozygous female parent?

b. Which locus is in the center?

c. Construct a genetic map.

A series of two-point mapping crosses indicated the following map distances involving the *b*, *h*, and *m* genes above and the additional linked genes *a*, *e*, *l*, and *s*:

$l–b$ = 3 mu

$l–e$ = 36 mu

$s–e$ = 21 mu

$s–m$ = 23 mu

$a–s$ = 33 mu

$a–e$ = 12 mu

$e–h$ = 27 mu

$e–m$ = 2 mu

d. Construct a genetic map for genes *a*, *b*, *e*, *h*, *l*, *m*, and *s* showing the correct sequences and distances.

21. In a certain species of plant, the following biochemical pathway is responsible for producing flower color variation:

$$AA \text{ or } Aa \qquad BB \text{ or } Bb \qquad DD \text{ or } Dd$$

white $\longrightarrow$ yellow $\longrightarrow$ red $\longrightarrow$ purple

Recessive genes (*aa*, *bb*, *dd*) do not produce active enzymes.

The linkage relationships for the three loci in question (*A*, *B*, *D*) are diagrammed to the right.

Consider the following crosses:

P: *aabbdd*   ×   *AAbbdd*   *aabbDD*   ×   *AABBdd*
                    ↓   ↓
F$_1$:   *Aabbdd*   ×   *AaBbDd*
                                    ↓
                                  F$_2$

a. Give the flower color for each plant in the P and F$_1$ generations above.

b. For the F$_2$ generation, give the expected percentages of plants with white, yellow, red, and purple flowers.

Team Problems

22. In humans, two recessive genes for "color weakness" occur on the X chromosome: one gene causes red weakness, and the other causes green weakness. Pedigree analysis identifies four types of families where the coupling/repulsion arrangement of the genes in females with normal vision can be determined by the phenotype of their fathers:

   type 1: mother's father was both red and green weak

   type 2: mother's father was neither red nor green weak

   type 3: mother's father was green weak only

   type 4: mother's father was red weak only

The following data were collected from various family types:

   Of 19 sons in families of type 1, 2 have green weakness.

   Of 27 sons in families of type 2, 2 have green weakness, and 1 has red weakness.

   Of 50 sons in families of type 3, 25 have red weakness, and 20 have green weakness.

   Of 60 sons in families of type 4, 28 have red weakness, and 26 have green weakness.

a. Are the mothers in the type 3 families in coupling or repulsion for the pair of recessive color weakness genes?

b. Which family type shows the greatest amount of crossing-over between the color genes?

c. Considering the offspring of all 156 sons in the study, how many of them inherited from their mothers an X chromosome that had undergone a crossover between the red-weak and the green-weak gene loci?

d. What is the map distance between the loci for red and green weakness?

23. It was not known whether certain genes were linked until the following crosses were performed:

	CROSS 1		CROSS 2		CROSS 3	
P crosses	$aaBBDD \times AAbbdd$		$bbEEGG \times BBeegg$		$ddGGHH \times DDgghh$	
	↓		↓		↓	
F$_1$ testcrosses	$AaBbDd \times aabbdd$		$BbEeGg \times bbeegg$		$DdGgHh \times ddgghh$	
	↓		↓		↓	
F$_2$ phenotypes	A B D	124	B E G	73	D G H	4
and numbers	a b d	120	b e g	77	d g h	6
	A B d	122	B E g	76	D G h	75
	a b D	126	b e G	74	d g H	65
	A b d	128	B e g	176	D g h	401
	a B D	130	b E G	174	d G H	389
	A b D	127	B e G	179	D g H	35
	a B d	123	b E g	171	d G h	25
totals		1,000		1,000		1,000

Answer the following questions individually for crosses 1, 2, and 3:

a. Which genes appear to be linked?

b. How many linkage groups are present?

c. Draw the genetic maps.

For the combined genetic data from all three crosses:

d. Draw a combined genetic map.

e. Give the total number of linkage groups, including all six gene loci, and briefly explain what these data tell us about the number of pairs of chromosomes in this species of organism.

Solutions

1. a. For the combined data, the chi-square value for the 1:1:1:1 ratio expected from independent assortment is 0.18, which is nonsignificant at 3 degrees of freedom.

   b. For data from plant 1, the chi-square value is 128.32; for data from plant 2, the chi-square value is 129.16. Both these yield P << 0.05 for 3 degrees of freedom. Thus, neither cross supports the hypothesis of independent assortment.

   c. Because the genes were in the coupling phase in plant 1 and in the repulsion phase in plant 2, the combination of data from the two plants *appears* to indicate independent assortment. This points out the danger of lumping together data from separate experiments.

2. Choice a, either cataract or polydactyly, is correct. Because he received cataract from one parent and polydactyly from the other, he must be in the repulsion phase of linkage. As his homologs separate at meiosis, each gamete would get one or the other chromosome. A crossover would be required to produce a chromosome with both or neither.

3. a. coupling $cvsn/+^{cv}+^{sn}$

   b. 2.1 mu. The total number of gametes is 5,000, of which 105 are recombinant. So, 105/5000 = 0.021.

   c. 21.0 (18.9 + 2.1 = 21.0)

4. a. $sr\ +^{e}$ 46%

         $+^{sr}\ e$ 46%

         $+^{sr}\ +^{e}$ 4%

         $sr\ e$ 4%

     The distance between the e and sr loci is $70 - 62 = 8$ mu = .08 recombination frequency.

   b.   50% wild type, 25% stripe, and 25% ebony, because there is no crossing-over in *Drosophila melanogaster* males and the repulsion phase males will produce 1/2 $sr\ +^{e}$ and 1/2 $+^{sr}\ e$ gametes. Use a Punnett square to determine the answer.

	SPERM	
EGGS	**0.5 $sr\ +^{e}$**	**0.5 $+^{sr}\ e$**
0.46 $sr\ +^{e}$	0.23 stripe	0.23 normal
0.46 $+^{sr}\ e$	0.23 normal	0.23 ebony
0.04 $sr\ e$	0.02 stripe	0.02 ebony
0.04 $+^{sr}\ +^{e}$	0.02 normal	0.02 normal

5. a.   r

   b.   q

   c.   m

   d.   n

To determine the center gene locus, first define both the parental and double-crossover classes by writing down the given most frequent and least frequent classes and their complements. For example, in a, the parental class phenotypes are r L E and R l e, and the double crossover class phenotypes are R L E and r l e. Double crossing-over changes the linkage phase of the r alleles. For c and d, it helps to write the alleles in the same order before proceeding. For example, in d, the parental class phenotypes are p X n and P x N, and the double-crossover class phenotypes are P x n and p X N.

6. a.   The males have pink eyes but neither curled wings nor prickly bristles because *Pr* and *Cu* are dominant.

   b.   Comparison of the parental ($p^{+}\ Cu\ Pr^{+}$ and $p\ Cu^{+}\ Pr$) and double-crossover ($p^{+}\ Cu^{+}\ Pr^{+}$ and $p\ Cu\ Pr$) classes indicates that the gene order is <u>p region 1 Cu region 2 Pr</u>.

   c.   <u>p 2 mu Cu 19.25 mu Pr</u>

The parental class ($p^{+}\ Cu\ Pr^{+}$ and $p\ Cu^{+}\ Pr$) is 79%; the single region 1 crossovers ($p\ Cu\ Pr^{+}$ and $p^{+}\ Cu^{+}\ Pr$) are 1.75%; the single region 2 crossovers ($p\ Cu\ Pr$ and $p\ Cu^{+}\ Pr^{+}$) are 19%; and the double crossovers ($p^{+}\ Cu^{+}\ Pr^{+}$ and $p\ Cu\ Pr$) are 0.25%. Therefore, the region 1 map distance is 1.75% + 0.25% = 2 mu, and the region 2 map distance is 19% + 0.25% = 19.25 mu.

   d.   0.64. The expected percentage of double crossing-over is $0.02 \times 0.1925 = 0.39\%$ The coefficient of coincidence = 0.25/0.39 = 0.64.

   e.   0.36. $1 - 0.64 = 0.36$.

7. a. $r\ Q/R\ q\ ;\ S/s$. The females are in the repulsion phase for the r and q loci, as can be seen by the parental classes. The s locus is either unlinked (in a different chromosome) or syntenic (> 50 mu from r or s on the same chromosome). There are two parental and two recombinant classes, and S and s are distributed equally between them, indicating independent assortment.

   b.   <u>r 32 mu q</u>   ; s

For r and q, there are a total of 320 recombinant progeny.

8. For no interference,

   a. 42

   b. 4,118

   c. 382

   d. 458

   e. With a coefficient of interference of 0.7,

       a) 13

       b) 4,088

       c) 412

       d) 487

   For this cross, the vermilion progeny are half the double-crossover class, the scalloped progeny are half the parental class, the wild-type progeny are half the single crossovers in region 2, and the sable progeny are half the single crossovers in region 1.

   In both cases, the four classes are determined by calculating the number of double-crossover progeny and subtracting them from each single-crossover class. For no interference:

   double crossovers = 0.1 × 0.085 = 0.85%
   single crossovers in region 1 = 10% – 0.85% = 9.15%
   single crossovers in region 2 = 8.5% – 0.85% = 7.65%
   parental types = 100% – 0.85% – 9.15% – 7.65% = 82.35%

   Thus, a vermilion = 0.0085 × 1/2 × 10,000 = 42.5 = 43.

   To compensate for interference, the double-crossover class is multiplied by the coefficient of coincidence.

   double crossovers = 0.1 × 0.85 × 0.3 = 0.26%
   single crossovers in region 1 = 10% – 0.26% = 9.74%
   single crossovers in region 2 = 8.5% – 0.26% = 8.24%
   parental types = 100% – 0.26% – 9.74% – 8.24% = 81.76%

9. 40.95% 2 cm, 9.05% 4 cm, 9.05% 6 cm, 40.95% 8 cm. Only the heterozygote can donate active alleles, and the number of active alleles donated depends on the genotypes of the eight classes of progeny.

CLASSES		FREQUENCIES		ACTIVE GENES	HEIGHT (cm)
double crossover	*AbC*	0.09 × 0.1 = 0.9%	0.45%	2	6
	aBc		0.45%	1	4
single crossover, region 1	*Abc*	9% – 0.9% = 8.1%	4.05%	1	6
	aBC	4.05%		2	6
single crossover, region 2	*ABc*	10% – 0.9% = 9.1%	4.55%	2	6
	abC		4.55%	1	4
parental	*ABC*	100% – 0.9% – 8.1%	40.95%	3	8
	abc	– 9.1% = 81.9%	40.95%	0	2

10. By the process of correlating the presence or absence of a specific chromosome with the presence of a specific enzyme activity, the following assignments may be made:

ENZYMES	CHROMOSOMES
a	17
b	2
c	4
d	20
e	17

11. Chromosome 8. The faster migrating band is found in all three subclones. Because hybrid cells retain the mouse chromosomes, we assume that this band is produced by the structural locus in the mouse genome. The slower migrating band is produced by the human structural locus, and its absence in subclone 3 indicates that the gene is in human chromosome 8.

12. 0.7125 *A–B–*, 0.2125 *aabb*, 0.0375 *A–bb*, and 0.0375 *aaB–*. Remember, no crossing-over occurs in *Drosophila* males.

13. 0.503 *F–S–*, 0.244 *F–ss*, 0.244 *ffS–*, and 0.005 *ffss*. In mice, crossing-over occurs in both sexes.

14. a. males and females = 0.425 *AB*, 0.425 *ab*, 0.075 *Ab*, and 0.075 *aB*
    b. males and females = 0.425 *Ab*, 0.425 *aB*, 0.075 *ab*, and 0.075 *AB*
    c. females = 100% *AB*; males = same as in part a
    d. females = 50% *Ab*, 50% *AB*; males = same as in part a
    e. females = 50% *Ab*, 50% *AB*; males = same as in part b

15. a. *b* and *c* are linked
    b. 10 mu (10%)

16. The classes with the smallest numbers are the offspring receiving double-crossover chromosomes. The gene in these classes that differs in *cis/trans* arrangement from the noncrossover offspring (the largest categories) is the gene in the middle of the sequence. So the eye color gene locus (*w*) is central, and the sequence is _y w ec_. With the sequence established, the single crossovers can be identified and the map distances calculated. The map is:

y	1.5 mu	*w*	4.1 mu	*ec*

17. The rate of double crossovers, producing BDg and bdG types is .06; the rate of single crossovers between the B and D loci, yielding BDG and bdg types is .14; the rate of single crossovers between the D and G loci, yielding BdG and bDg types is .24; and the rest (.56) are noncrossovers.
    a. 0.49
    b. 70 of BDG
       280 of Bdg
       30 of BDg
       120 of bDg
       120 of BdG
       30 of bdG
       280 of bDG
       70 of bdg

18. a. The hair length locus (S, s alleles)

    b. ___B _____10 mu_____S_ 2 mu _H___

19. a. No, because 50% recombinant offspring result from the testcross.

    b. no

    c. Yes, because only 20% recombinant offspring result from the testcross.

    d. two linkage groups

    e. ___A _____20 mu_____G_____ ___D____

    f. _C_ 7 mu _A_ 5 mu _E_____15 mu_____G___

    _D_____10 mu_____F_____20 mu_____B___

    g. two linkage groups

20. a. $b\ m^+\ h^+$ and $b^+\ m\ h$, by determining the parental classes

    b. m, by comparison of the parental and double-crossover classes

    c. ___b_____35 mu_____m_____25 mu_____h___

    d. _l_ 3 mu _b_ 12 mu __s_____21 mu_____e_ 2 mu _m_ 10 mu _a_____15 mu_____h___

    Use the calculated and given map distances to determine a sequence that makes additive mathematical sense. Because gene distances are roughly additive, the distances b to m (12 + 21 + 2 = 35) and m to h (10 + 15 = 25) are the same as in part c above.

21. a. P: $aabbdd$ × $AAbbdd$ $aabbDD$ × $AABBdd$
              white   **yellow**   **white**   **red**

                         ↓   ↓

        F$_1$: $Aabbdd$ $AaBbDd$
                        yellow   **purple**

    b. white = 25%

       yellow = 37.5%

       red = 28.125%

       purple = 9.375%

22. a. In family types 1 and 2, the mother is in the coupling phase, and sons with either red or green weakness, but not both, represent *recombinant types*.

    a) In families of type 1, 2 of 19 sons are recombinant.

    b) In families of type 2, 3 of 27 sons are recombinant.

    In family types 3 and 4, the mother is in the repulsion phase, and sons with either red or green weakness, but not both, represent *parental types*.

    a) In families of class 3, 45 of 50 sons are parental types, and the remaining 5 are recombinant types.

    b) In families of class 4, 54 of 60 sons are parental types, and the remaining 6 are recombinant types.

b. crossover rates:

2/19 = 0.105 for family 1 types

3/27 = 0.111 for family 2 types

5/50 = 0.100 for family 3 types

6/60 = 0.100 for family 4 types

c. A total of 16 of 156 sons are recombinant types.

d. 16/156 yields a frequency of 0.103 (10.3%). So there are 10.3 map units of genetic distance between the gene loci.

23. | **CROSS 1** | **CROSS 2** | **CROSS 3** |
|---|---|---|
| a. none | B and E | all 3 |
| b. 3 | 2 | 1 |
| c. $\underline{A \ B \ D}$ | $\underline{B \quad 30\,mu \quad E} \ \ G$ | $G \quad 15\,mu \quad D \quad 7\,mu \ H$ |

d. combined genetic map:

$\underline{B \quad 30\,mu \quad E} \quad \underline{G \ \ 15\,mu \ \ D \ \ 7\,mu\,H} \quad \underline{A}$

e. Three linkage groups are present indicating three pairs of chromosomes: one pair with the *A* locus, a second pair with the *G, D,* and *H* loci, and a third pair with the *B* and *E* loci.

8
Chromosomal Aberrations

Problems

1. Consider a species with a diploid ($2n$) number of 20 chromosomes. How many chromosomes would be found in a:
    a. trisomic body cell?
    b. nullisomic body cell?
    c. monoploid body cell?
    d. disomic gamete?
    e. nullisomic gamete?

2. How many different trisomics could be formed in a plant with a diploid number of 40?

3. In *Drosophila melanogaster*, the recessive gene eyeless (*ey*) is located on the small fourth chromosome. Flies monosomic and trisomic for this chromosome are viable, and disomic and nullisomic sperm are fully viable. Give the results of the following crosses for both the eye phenotype and chromosome constitution:
    a. *ey* male × *ey$^+$ ey* female
    b. *ey$^+$ ey$^+$ ey* male × *ey$^+$ ey* female

4. In maize, red aleurone (*R*) is dominant over colorless (*r*). Disomic ovules are functional, but disomic pollen are not. Give the results of the following crosses for both the aleurone phenotype and chromosome constitution:
    a. *R r r* female × *R r* male
    b. *R r r* male × *R r* female

5. The normal sequence of genes and centromere for a particular chromosome is:
    1   2   3   4   •   5   6   7   8

    a. Determine the chromosome aberration in each of the following chromosomes:

a)	1	2	3	6	5	•	4	7	8
b)	1	2	3	4	•	5	6	8	
c)	3	2	1	4	•	5	6	7	8
d)	1	2	2	3	4	•	5	6	7	8

    b. Use diagrams to show how each of these aberrant chromosomes would pair with the normal chromosome.

6. In *Drosophila*, a male with the recessive genotype *rrssttuu* is mated to a female from a strain homozygous for the dominant wild-type alleles of these genes, and they produce progeny half of which are wild type and half of which show the recessive traits corresponding to *s* and *t*. These genes are linked in chromosome 2 in the order *r-s-t-u*. Explain these results.

7. Seven bands in a *Drosophila* salivary gland chromosome are shown below, along with the extent of six duplications (i.e., duplication 1 repeats bands 1, 2, and 3).

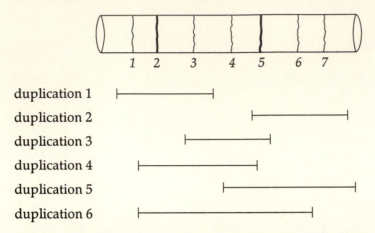

Flies carrying individual duplications were tested for the amounts of two different enzymes. Normal diploid flies produce 70 units of enzyme A and 120 units of enzyme B. However, in strains carrying the duplications, the following amounts of enzymes were found:

	DUPLICATIONS					
ENZYME	1	2	3	4	5	6
A	103	69	72	106	70	104
B	122	179	182	118	180	177

Which chromosome band most likely carries the gene coding for enzyme A? For enzyme B?

8. Hypothetically, several closely linked genes in fruit flies are located in a region of a polytene chromosome containing bands numbered 1 through 14. Homozygous wild-type males (*QQRRSSTT*) are irradiated and are testcrossed with homozygous recessive females (*qqrrsstt*). In eight such matings, the following patterns of pseudodominance were observed, and the bands deleted from the irradiated chromosomes were determined to be as follows:

EXPERIMENT	PATTERN OF PSEUDODOMINANCE	BANDS DELETED
1	q R S T	1–2
2	Q R s T	6–9
3	Q R S T	11–14
4	Q R s T	4–8
5	Q r s T	3–7
6	Q R S t	10–14
7	q r S T	1–5
8	Q R S t	8–14

Determine the exact band locations for genes *Q, R, S,* and *T*.

9. In fruit flies, Bar (B) eyes is caused by a tandem duplication of a segment of the X chromosome. Ultrabar (B^U) females have three doses of the segment in at least one X. Homozygous Bar females produce normal (B^+) or ultrabar male offspring at the rate of approximately 1 in 1,600.

   a. Show diagrammatically how these reversions occur.

   b. In 1925, Sturtevant crossed $f^+\,B\,fu^+/f\,B\,fu$ females with f B fu/Y males (f = forked bristles, fu = fused wing veins). He classified 18,999 offspring as shown below:

BAR				NORMAL			ULTRABAR
$f^+\,fu^+$	$f\,fu$	$f^+\,fu$	$f\,fu^+$	$f^+\,fu^+$	$f^+\,fu$	$f\,fu^+$	$f\,fu^+$
10,631	7,909	187	264	1	2	3	2

   Illustrate how normal-eyed and ultrabar could originate from homozygous Bar females by single meiotic events.

10. Different sequences of genes on chromosome 9 exist in four strains of corn. These sequences are:

strain 1:	D^+	yg_2	bz	sh_1	c	bp	wx
strain 2:	D^+	bp	c	yg_2	bz	sh_1	wx
strain 3:	D^+	yg_2	c	bp	bz	sh_1	wx
strain 4:	D^+	yg_2	c	sh_1	bz	bp	wx

   If strain 1 is the ancestral strain, indicate the sequence in which these four strains evolved, including the inversions that have occurred.

11. A female fruit fly is homozygous for the following sequence of genes in chromosome 3: $A\,B\,D\,E\,F\,G \cdot H\,M$. A male fly is homozygous for the sequence $a\,d\,b\,e\,f\,h \cdot g\,m$. These flies were mated to produce offspring.

    a. Draw the third chromosomes as they would pair at the tetrad stage of prophase I of meiosis in the F_1 females, including genes and centromeres.

    b. Number the chromatid ends in your diagram for part a 1 through 4 down the left side and 5 through 8 down the right side. Draw the chromosomes during anaphase I if crossing-over occurred between genes B and D of chromatids 1 and 3 and between gene G and the centromere of chromatids 2 and 4.

    c. In another F_1 female, draw the chromosomes during anaphase I if crossing-over occurred between gene G and the centromere of chromatids 1 and 3.

12. A species of fruit fly is differentiated into six varieties correlated with differences in the banding sequences of one of its polytene salivary gland chromosomes. Eight bands on the chromosomes are designated a through h. If each race is related to another by a single inversion difference, devise a general evolutionary scheme for the six races if their banding sequences are:

    race 1: a b c d e f g h

    race 2: a e d h g f b c

    race 3: a b c f e d g h

    race 4: b c d e a f g h

    race 5: a e d c b f g h

    race 6: a f g h d e b c

13. In *Drosophila*, the locus for the white eye color (*w*) gene is near the tip of the X chromosome. A normal-eyed w^+ Y male is X-rayed and then mated with *ww* females. One of their male offspring has normal eye color (w^+), an unexpected result. This unusual male is mated with *ww* females, and they produce 100 progeny, of which half are white-eyed females and half are wild-type males. Explain these results.

14. The following diagram represents a pair of homologous chromosomes in an individual:

The upper chromosome contains the inversion. From this individual, a rare offspring inherited the following genes: *A B D e f G H*.

  a. Draw a diagram indicating what happened during prophase I of meiosis to account for this offspring.

  b. Is the gene sequence in this offspring in the normal or inverted sequence?

15. Much chromosomal polymorphism exists among the various species of mantis in the genus *Ameles*. Studies by Wahrman and O'Brien in 1956 produced the following data for several species:

SPECIES	MALE DIPLOID CHROMOSOME NUMBER	NUMBERS OF METACENTRIC	ACROCENTRIC	AVERAGE DNA PER SPERMATID
1	19	11	8	2.07
2	20	10	10	2.11
3	21	9	12	1.96
4	27	3	24	1.84
5	28	2	26	1.71

Explain the differences in chromosome number and type among these species, considering that the DNA amount is quite similar for all species.

16. A region of one of the polytene chromosomes of *Drosophila* has 13 bands. A large study of naturally occurring chromosomal aberrations has discovered five different deletions and six different duplications. Mating females with one or another of the deletions with males homozygous recessive for genes *aabbccddeeff* yielded the data in the first of the following tables. Measuring the amount of enzyme produced by gene *G* in females heterozygous for one or another of the duplications and a normal chromosome yielded the data found in the second table. Normal diploid flies produce about 100 units of enzyme "G." Indicate the exact chromosomal band locations of all seven of the gene loci.

DELETION	BANDS DELETED	PHENOTYPES OF FEMALES HETEROZYGOUS FOR DELETION AND GENE					
I	1–3	a	b	c	D	E	F
II	2–3	a	B	c	D	E	F
III	2–4	a	B	c	D	e	F
IV	3–5	A	B	c	d	e	F
V	4–6	A	B	C	d	e	f

DUPLICATION	BANDS DUPLICATED	AMOUNT OF ENZYME "G" PRODUCED BY FEMALE HETEROZYGOUS FOR DUPLICATION
VI	7–10	143 units
VII	8–13	152 units
VIII	9–10	160 units
IX	9–11	148 units
X	9–13	155 units
XI	10–13	98 units

Challenge Problems

17. A region of a polytene chromosome in *Drosophila* has 15 bands (numbered 1–15). A series of X-ray treatments (experiments 1–10) produced different deletions (I–X) in this region. The bands deleted in each case are indicated below. In each experiment, a male homozygous dominant for two linked gene pairs (*EEHH*) was irradiated and then mated to females homozygous recessive *eehh*. Thus, *eh/eh* females were mated with *EH/EH* X-rayed males. Heterozygous female offspring (*EH/eh*) produced the patterns of "pseudodominance" indicated below. For instance, a heterozygous female offspring carrying deletion I is pseudodominant for trait h, but not for E.

EXPERIMENT	DELETION INDUCED	BANDS DELETED	PHENOTYPES OF HETEROZYGOUS FEMALES FROM THE CROSS	
1	I	8–15	E	h
2	II	8–12	E	h
3	III	1–10	e	H
4	IV	3–5	e	H
5	V	4–10	E	H
6	VI	6–11	E	H
7	VII	1–6	e	H
8	VIII	11–15	E	h
9	IX	1–9	e	H
10	X	7–15	E	h

18. Two related varieties of fruit flies (varieties 1 and 2) were crossed in the laboratory, and the polytene chromosomes of the resulting hybrid larvae were examined. A diagram of a portion of chromosome 3 from a hybrid is given below:

    a. On the basis of the inversion loops present, give the correct sequence of genes in the chromosomes from both varieties.

    b. Three other closely related varieties (3, 4, and 5) have gene sequences as follows:

    variety 3:   1 2 3 4 5 6 13 12 11 10 9 8 7 14 15

    variety 4:   1 2 3 4 5 6 7 8 9 10 11 12 13 14 15

    variety 5:   1 2 3 4 9 10 11 12 13 6 5 8 7 14 15

    If variety 4 is the oldest in the group, indicate the evolutionary relationships of the five varieties by means of an arrow diagram.

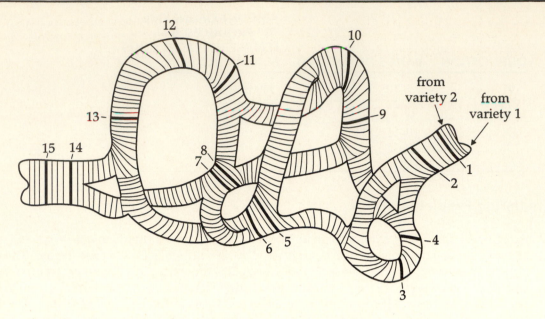

Team Problems

19. Seven bands in a *Drosophila* salivary gland chromosome are shown below, along with the extent of five deletions (i.e., deletion 1 eliminates bands 3, 4, and 5).

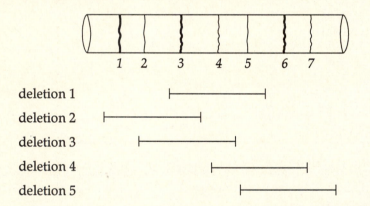

Recessive alleles *a, b, c, d, e, f,* and *g* are in the region, and these deletions may be used to determine the gene sequence. Females heterozygous for the seven gene pairs and heterozygous for one of the deletions show the following patterns of pseudodominance:

DELETION TYPE IN THE FEMALES	PATTERN OF PSEUDODOMINANCE FOR GENES						
	a	*b*	*c*	*d*	*e*	*f*	*g*
1	a	b	C	D	e	F	G
2	a	B	c	D	E	F	g
3	a	B	C	D	e	F	g
4	A	b	C	D	e	f	G
5	A	b	C	d	E	f	G

Use these data to determine the salivary gland chromosome band position of each gene.

20. The diagrams below indicate the diploid gene orders and chromosome morphologies for six closely related theoretical species of insect. Describe in detail how these species probably evolved from each other, assuming that species C is the common ancestor.

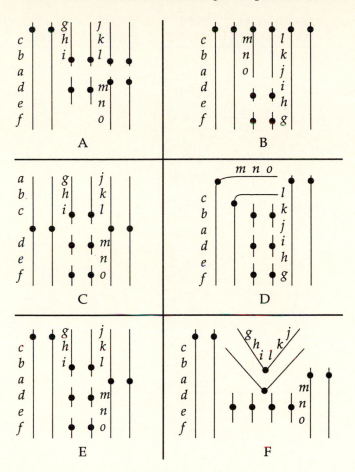

Solutions

1. a. $21 = 2n + 1$
   b. $18 = 2n - 2$
   c. $10 = 1n$
   d. $11 = 1n + 1$
   e. $9 = 1n - 1$

2. 20, one for each chromosome type

3. a. 1/4 normal-eyed disomic, 1/4 eyeless disomic, 1/4 normal-eyed monosomic, 1/4 eyeless monosomic
   b. 3/12 normal-eyed trisomic, 5/12 normal-eyed disomic, 1/12 normal-eyed monosomic, 1/12 eyeless disomic, 2/12 eyeless monosomic

The answers are determined by using the forked-line method for gametes to show the various possible fertilizations. In determining gametes, it helps to label homologs carrying the same allele in determining all possible segregations. For example, for part b:

TYPES OF SPERM	TYPES OF EGGS	FREQUENCIES OF OFFSPRING	TYPE OF EYE	NUMBER OF FOURTH CHROMOSOMES
1/6 ey-1 ey^+	↗ 1/2 ey^+ →	1/12	normal	3
	↘ 1/2 O →	1/12	normal	2
1/6 ey-2	↗ 1/2 ey^+ →	1/12	normal	2
	↘ 1/2 O →	1/12	eyeless	1
1/6 ey-2 ey^+	↗ 1/2 ey^+ →	1/12	normal	3
	↘ 1/2 O →	1/12	normal	2
1/6 ey-1	↗ 1/2 ey^+ →	1/12	normal	2
	↘ 1/2 O →	1/12	eyeless	1
1/6 ey-1 ey-2	↗ 1/2 ey^+ →	1/12	normal	3
	↘ 1/2 O →	1/12	eyeless	2
1/6 ey^+	↗ 1/2 ey^+ →	1/12	normal	2
	↘ 1/2 O →	1/12	normal	1

4. a. 5/12 red trisomic, 4/12 red disomic, 2/12 colorless disomic, 1/12 colorless trisomic. The functional female ovules are 2/6 R r (R r^1 and R r^2), 2/6 r (r^1 and r^2), 1/6 r^1 r^2, and 1/6 R; and the pollen are 1/2 R, 1/2 r.

    b. All progeny are disomic: 4/6 red and 2/6 colorless. The functional pollen are 2/3 r (r^1 and r^2) and 1/3 R, and the ovules are 1/2 R and 1/2 r.

5. a. Types of chromosome aberrations:

    a) a pericentric inversion involving region <u>4 • 5 6</u>

    b) an interstitial deletion involving region <u>7</u>

    c) a paracentric inversion involving region <u>1 2 3</u>

    d) a tandem duplication involving region <u>2</u>

    b.

The normal chromosome is at the top in each diagram.

6. The wild-type female parent is heterozygous for a deletion in chromosome 2 involving the region containing the s and t loci. Her genotype is $r^+s^+t^+u^+/r^+u^+$. Half her progeny receive chromosomes $r^+s^+t^+u^+/rstu$ and are wild type in phenotype, and half receive chromosomes $r^+ u^+/rstu$ and express the s and t traits.

7. By correlating the amount of enzyme with the duplication regions, the following map assignments may be made: Band 2 is present in 3 doses only in those flies with $3x$ amount of enzyme A ($x = 35$ units of A). Thus, there are 3 doses of the gene for enzyme A at band 2. Band 5 is present in 3 doses only in those flies with $3x$ amount of enzyme B ($x = 60$ units of A). Thus, there are 3 doses of the gene for enzyme B at band 5.

8. Locus Q is in band 1 or 2. Locus R is in band 3. Locus S is in band 6 or 7. Locus T is in band 10.

9. a.

b.

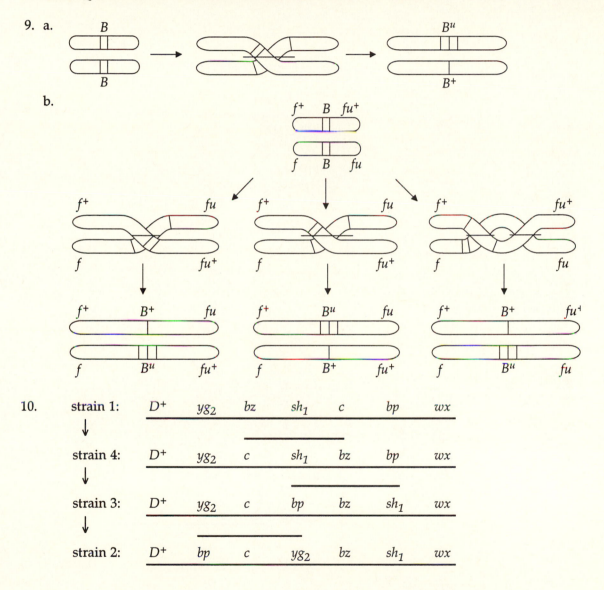

10.

strain 1:	D^+	yg_2	bz	sh_1	c	bp	wx
strain 4:	D^+	yg_2	c	sh_1	bz	bp	wx
strain 3:	D^+	yg_2	c	bp	bz	sh_1	wx
strain 2:	D^+	bp	c	yg_2	bz	sh_1	wx

11. a.

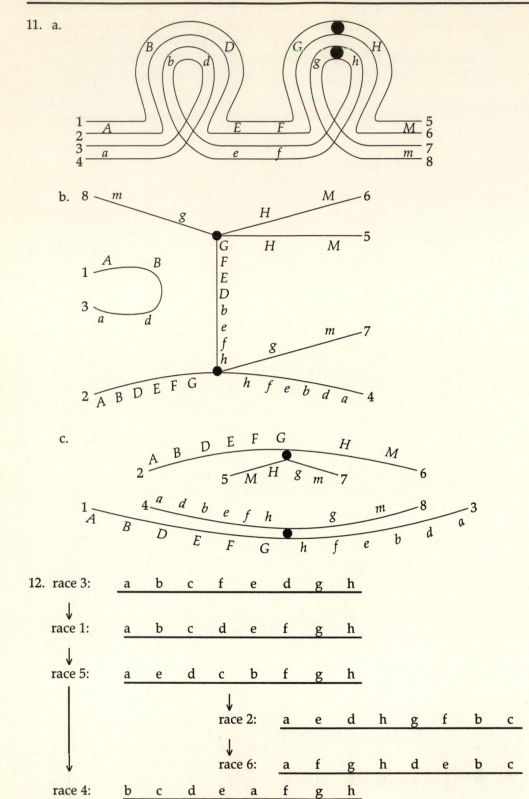

b.

c.

12. race 3:   a   b   c   f   e   d   g   h

   ↓

race 1:   a   b   c   d   e   f   g   h

   ↓

race 5:   a   e   d   c   b   f   g   h

                    ↓

                      race 2:   a   e   d   h   g   f   b   c

                           ↓

                      race 6:   a   f   g   h   d   e   b   c

race 4:   b   c   d   e   a   f   g   h

13. The most likely explanation is that X-ray treatment of the w^+ Y male caused a translocation of the tip of one X chromosome containing the w^+ locus onto the Y chromosome and this w^+-Y was inherited by the normal-eyed male offspring, his genotype being w-X, w^+-Y. When he was mated to ww females, the progeny would receive either ww and be white-eyed females or w w^+-Y and be wild-type males.

14. a. During prophase I of meiosis, a double crossover (between D–e and between f–G) within the inversion loop occurred:

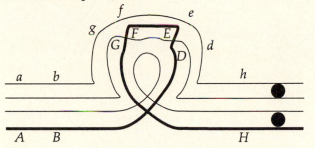

    b. The gene sequence is in the normal (noninverted) sequence.

15. The total number of chromosome *arms* remains the same because each species has 30 arms per diploid cell. For instance, species 1 has 11 metacentric (22 arms) and 8 acrocentric (8 arms) chromosomes, totaling 19 chromosomes (30 arms). So these species are related through a series of centric fusions (if species 5 is ancestral) or breakages at the centromeres (if species 1 is ancestral).

16. A = band 2
    B = band 1
    C = band 3
    D = band 5
    E = band 4
    F = band 6
    G = band 9

17. Locus E is in band 3, and locus H is in band 12. For locus E, use the following reasoning: In experiment 1, bands 8–15 are deleted, but E is still present; thus, E is in bands 1–7. In experiment 4, bands 3–5 are deleted, and E is deleted; thus, E is in bands 3–5. In experiment 5, bands 4–10 are deleted, but E is still present; thus, E is in band 3. The reasoning used for locus H is similar: In experiment 2, bands 8–12 are deleted, and H is deleted; thus, H is in bands 8–12. In experiment 3, bands 1–10 are deleted, but H is still present; thus, H is in bands 11–12. In experiment 6, bands 6–11 are missing, but H is still present; thus, H is in band 12.

18.

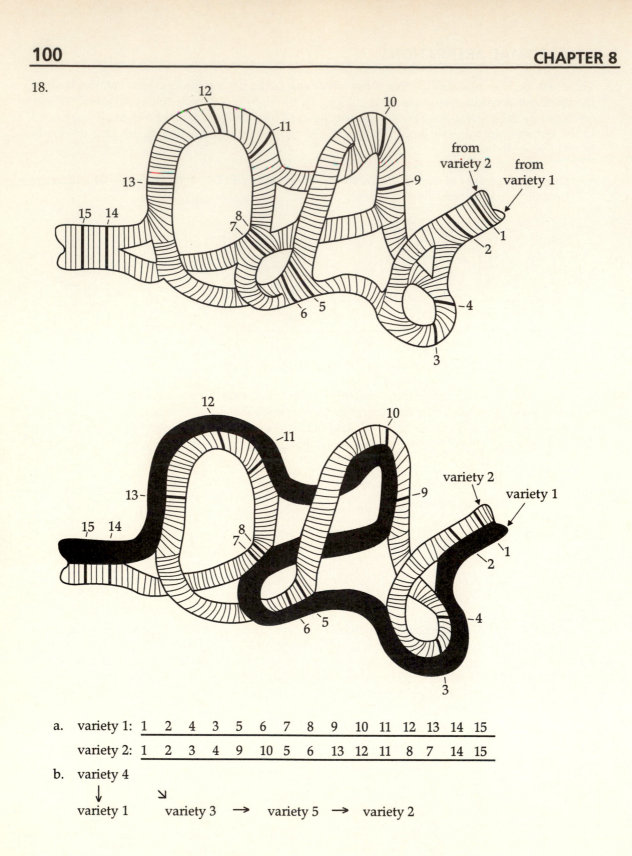

a.   variety 1: <u>1   2   4   3   5   6   7   8   9   10 11 12 13 14 15</u>

      variety 2: <u>1   2   3   4   9   10 5   6   13 12 11 8   7   14 15</u>

b.   variety 4
            ↓   ↘
      variety 1   variety 3 → variety 5 → variety 2

19. By correlating the pseudodominance data with the deletion regions, the following map assignments may be made:

> Band 1 is deleted only in flies with phenotype c.
>
> Band 2 is deleted only in flies with phenotype g.
>
> Band 3 is deleted only in flies with phenotype a.
>
> Band 4 is deleted only in flies with phenotype e.
>
> Band 5 is deleted only in flies with phenotype b.
>
> Band 6 is deleted only in flies with phenotype f.
>
> Band 7 is deleted only in flies with phenotype d.

20.

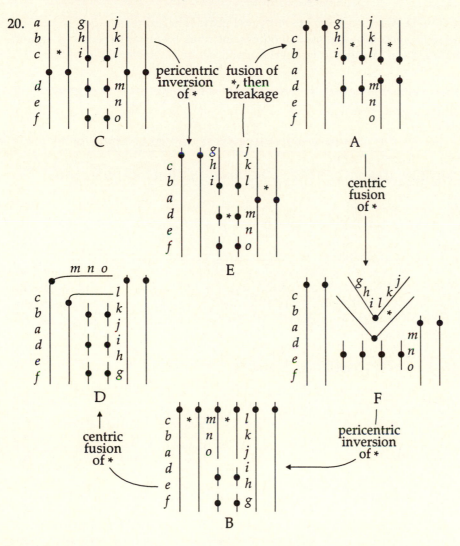

9
Metabolic Pathways

Problems

1. Four nutritional mutants in *Neurospora*, arg^1 to arg^4, will grow if arginine is added to minimal medium. Some of these mutants also will grow if other substrates chemically related to arginine are added to minimal medium. The growth patterns are shown in the table (+ = growth, 0 = no growth).
   a. What is the normal sequence of events in the synthesis of arginine?
   b. What product accumulates in each mutant?

		SUPPLEMENTS ADDED TO MINIMAL MEDIUM			
MUTANTS	CONTROL	GLUTAMIC SEMIALDEHYDE	ARGININE	CITRULLINE	ORNITHINE
3	0	+	+	+	+
4	0	0	+	+	+
2	0	0	+	+	0
1	0	0	+	0	0

2. Six different mutant strains of *Neurospora* are unable to grow on minimal medium unless it is supplemented by one or more of the substrates A through F. In the following table, mutant growth is indicated by a "+" and no growth by a "0." Propose a metabolic pathway consistent with these data, indicating at which step each gene acts and the positions of each substrate.

	SUPPLEMENTS ADDED TO MINIMAL MEDIUM						PRODUCTS
STRAIN	A	B	C	D	E	F	ACCUMULATED
1	0	+	0	+	+	0	A
2	0	0	0	+	0	0	A, B
3	+	0	0	+	0	+	B, C
4	+	0	0	+	0	0	B, C, F
5	0	+	0	+	0	0	A, E
6	+	0	+	+	0	0	B, F

3. Seven nutritional mutants that responded to the addition of certain supplements to minimal medium by growth (+) or no growth (0) were isolated from *Neurospora*. Given the following responses for the seven mutants (1 through 7 in the table) and the compounds accumulated by these mutants, diagram the metabolic pathway that exists in the normal strain consistent with all the data.

| | SUPPLEMENTS ADDED TO MINIMAL MEDIUM | | | | | | | | | | | | PRODUCTS |
STRAIN	A	D	E	F	H	R	T	U	D&T	F&U	B&P	F&H	ACCUMULATED
1	0	0	+	0	0	+	0	0	+	0	0	0	B, F
2	+	0	+	0	0	+	0	+	+	+	+	0	F, H
3	0	0	0	0	0	+	0	0	+	0	0	0	E
4	0	0	+	+	0	+	0	0	+	+	+	+	B, ?*
5	+	0	+	0	+	+	0	+	+	+	+	+	F, ?
6	0	0	0	0	0	0	0	0	+	0	0	0	R
7	+	0	+	0	0	+	0	0	+	0	+	0	F, U

*? = unidentified compound, probably an early precursor

4. Several nutritional mutant strains were isolated from wild-type *Neurospora*. These mutants responded to the addition of certain supplements in the culture medium by growth (+) or no growth (0). Diagram a biochemical pathway consistent with the following data:

| | GROWTH FACTORS | | | |
STRAIN	A	B	C	D
1	0	0	+	+
2	0	0	0	+
3	+	0	+	+
4	0	+	+	+

5. Several nutritional mutant strains were isolated from wild-type *Neurospora*. These mutants responded to the addition of certain supplements in the culture medium by growth (+) or no growth (0). Given the following responses for six single gene mutants (numbered 1–6), diagram a metabolic pathway consistent with all the data.

| | SUPPLEMENTS ADDED TO MINIMAL MEDIUM | | | | | | | | PRODUCTS |
MUTANT	B	D	I	M	N	O	P	U	X	ACCUMULATED
1	0	+	0	0	0	0	0	0	+	I, N
2	0	+	0	+	+	0	0	0	0	B, P, X
3	0	+	0	0	0	0	0	0	0	N, X
4	0	+	+	0	0	0	0	0	+	N, U
5	0	+	0	0	+	0	0	0	0	M, X
6	0	+	+	0	0	0	0	+	+	N, O

6. Several nutritional mutant strains were isolated from wild-type *Neurospora*. These mutants responded to the addition of certain supplements in the culture medium by growth (+) or no growth (0). Given the following responses for five gene mutants (numbered 1–5), diagram a metabolic pathway consistent with all the data.

| | SUPPLEMENTS ADDED TO MINIMAL MEDIUM | | | | |
MUTANT	A	B	C	D	E
1	0	+	0	+	0
2	0	+	+	+	0
3	0	+	0	0	0
4	+	+	0	0	0
5	+	+	0	0	+

Challenge Problems

7. Individuals homozygous for any of the nonallelic mutant genes *i*, *f*, *t*, *h*, and *g* cannot make compound W, which is necessary for growth. Experiments show that the compounds listed in column 1 of the following table accumulate in the mutant individuals listed in column 2. Growth occurs in the mutant individuals listed in column 3 when the column 1 compounds are fed to them. Growth does not occur in the mutant individuals listed in column 4 when the column 1 compounds are fed to them. Using the concepts Beadle and Tatum developed, explain these results in the form of a metabolic pathway.

	COLUMN			
EXPERIMENT	1	2	3	4
1	G	*g*	*i, f*	*h, t*
2	E	*t*	*h, f, g, i*	—
3	A	*i*	*f*	*h, t, g*
4	N	*h*	*g, f, i*	*t*

8. Five mutant strains (1 through 5) of *Salmonella* bacteria were isolated, each controlling a different step in a single biochemical pathway, the end product of which is compound A. Using the minimal medium petri dish growth patterns shown below and the compounds that accumulate in each case, determine the sequence of steps in the normal biochemical pathway producing compound A.

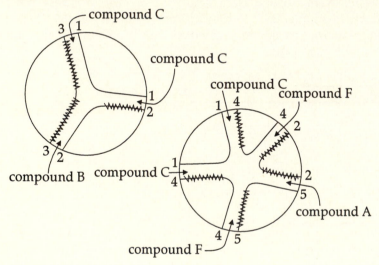

Team Problems

9. Suppose that the following metabolic pathway exists in *Neurospora:*

$$K \xrightarrow{1} \begin{array}{c} M \xrightarrow{2} R \xrightarrow{3} \begin{array}{c} S \\ W \xrightarrow{4} P \end{array} \\ H \end{array}$$

Based on this pathway, fill in the following growth table, using "+" for growth and "0" for no growth. Also list the expected accumulation products.

STRAIN	H	R	P	S&W	H&M	H&S&W	H&R	P&S	COMPOUNDS ACCUMULATED
				SUPPLEMENTS ADDED TO MINIMAL MEDIUM					
1									
2									
3									
4									
5*	0	+	0	+	0	+	+	+	H, L
6*	+	0	+	0	+	+	+	+	B, W

*Two new mutants (5 and 6) showed these growth patterns. Propose a new metabolic pathway showing the additional biochemical details indicated by these new mutants.

10. The following biochemical pathway for the synthesis of compound G was determined in *Salmonella* bacteria using five mutant strains (strains 1–5) and the compounds (A–F) accumulated by these mutant strains when they couldn't grow.

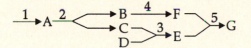

Below, on the diagram of a petri dish containing minimal medium and streaks of the various mutant bacterial strains, indicate where growth is expected to occur (use wavy lines). In the blank spaces, indicate which compounds will accumulate due to the strains that cannot grow.

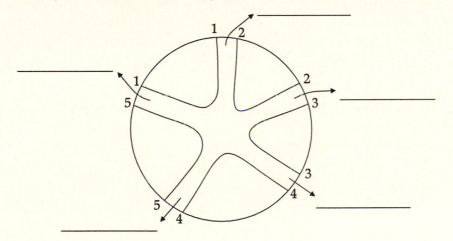

Solutions

1. a. According to the rules developed by Beadle and Tatum, the growth patterns of the various mutant strains give the following information:

    $$arg^3 \longrightarrow \begin{matrix} \text{GSA} \\ \text{arginine} \\ \text{citrulline} \,, \\ \text{ornithine} \end{matrix} \qquad \text{GSA} \xrightarrow{arg^4} \begin{matrix} \text{arginine} \\ \text{citrulline} \,, \\ \text{ornithine} \end{matrix}$$

    $$\begin{matrix} \text{GSA} \\ \text{ornithine} \end{matrix} \xrightarrow{arg^2} \begin{matrix} \text{arginine} \\ \text{citrulline} \,, \end{matrix}$$

    $$\begin{matrix} \text{GSA} \\ \text{citrulline} \\ \text{ornithine} \end{matrix} \xrightarrow{arg^1} \text{arginine}$$

    Therefore, the steps in the normal synthesis of arginine in *Neurospora* are:

    $$\xrightarrow{arg^3} \text{GSA} \xrightarrow{arg^4} \text{ornithine} \xrightarrow{arg^2} \text{citrulline} \xrightarrow{arg^1} \text{arginine}$$

    b. The expected accumulation products would be: unknown for arg^3, GSA for arg^4, ornithine for arg^2, and citrulline for arg^1.

2. Because more than one accumulation product occurs for some mutants, the metabolic pathway has several branches. The growth data may be interpreted as follows, listing only accumulation products before blocked steps:

1	2	3
A → B, D, E;	A&B → D;	B&C → A, D;

4	5	6
B&C&F → A, D;	A&E → B, D;	B&F → A, C, D

    One pathway that is consistent with these data:

    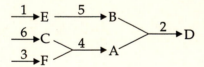

3. The growth data may be interpreted as follows:

$$B\&F \xrightarrow{1} E, R, (D+T); \qquad F\&H \xrightarrow{2} B, E, R, U, (D+T), (F+U), (B+F);$$

$$E \xrightarrow{3} R, (D+T); \qquad B\&? \xrightarrow{4} E, F, R, (D+T), (F+U), (B+F), (F+H);$$

$$F\&? \xrightarrow{5} B, E, H, R, U, (D+T), (F+U), (B+F), (F+H); \qquad R \xrightarrow{6} (D+T);$$

$$F\&U \xrightarrow{7} B, E, R, (D+T), (B+F)$$

One metabolic pathway consistent with these data:

4.

5.

6.

7. $-f \rightarrow A - i \rightarrow G - g \rightarrow N - h \rightarrow E - t \rightarrow W$

8. $-3 \rightarrow B - 2 \rightarrow A - 5 \rightarrow F - 4 \rightarrow C - 1 \rightarrow E$

9. Based on the metabolic pathway given, the following growth data would be anticipated:

| | SUPPLEMENTS ADDED TO MINIMAL MEDIUM | | | | | | | | COMPOUNDS |
STRAIN	H	R	P	S&W	H&M	H&S&W	H&R	P&S	ACCUMULATED
1	0	0	0	0	+	+	+	+	K
2	0	+	0	+	0	+	+	+	H, M
3	0	0	0	+	0	+	0	+	H, R
4	0	0	+	0	0	0	0	+	H, W

Growth data from the two newly discovered mutants may be interpreted as follows:

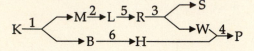

H&L $\xrightarrow{5}$ R, (S+W), (H+S), (H+R), (P+S);

B&W $\xrightarrow{6}$ H, P, (H+M), (H+S), (H+R), (P+S)

One metabolic pathway consistent with these data:

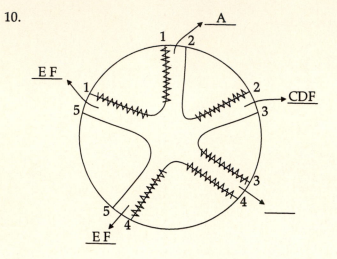

10.

10
Quantitative Inheritance

Problems

1. If three independently assorting gene pairs (*Aa, Bb, Cc*) determine height in a particular plant such that each active gene (in capital letters) adds 2 cm to a "base height" of 2 cm:

   a. Give the heights expected in the F_1 progeny of a cross between two homozygous stocks *AABBCC* (14 cm) × *aabbcc* (2 cm).

   b. Give the distribution of heights (phenotypes and frequencies) expected among the offspring of an $F_1 \times F_1$ cross.

   c. What proportion of the F_2 generation will have heights equal to the parental stocks?

   d. What proportion of the F_2 will breed true for the height shown by the F_1 plants?

2. What results are expected in problem 1 if, instead of acting additively, each active gene multiplied height by 2?

3. Skin color in humans, from very dark to very fair, may be controlled by four independently assorting pairs of genes with equal and additive effects (very dark = *AABBCCDD* and very fair = *aabbccdd*). From a mating between two intermediates (*AaBbCcDd*), what is the probability of producing:

   a. a very dark child?

   b. a very fair child?

   c. a child intermediate like the parents?

   d. a child darker by one color gene than the parents?

4. A cross between two inbred plants that had seeds with mean weights of 20 and 40 cg, respectively, produced an F_1 generation with seeds weighing an average of 30 cg each. An $F_1 \times F_1$ cross produced 1,000 plants: 3 had seeds weighing 20 cg, 5 had seeds weighing 40 cg, and the others produced seeds with weights varying between these extremes. How many gene pairs are involved in determining seed weight differences in these plants?

5. In a cross between a heavy and a light strain of rabbits, the F_1s are intermediate in weight between the parents and show little variation. Among 2,025 F_2 individuals, 8 are the same weight as the lighter parental strain, and 7 are the same weight as the heavier parental strain. Assuming weight to be governed by additive alleles, how many pairs of genes affect weight in these strains?

6. Clausen and Hiesey made crosses between two California races of *Potentilla glandulosa*. One was a coastal population active in growth all year round, and the other was an alpine population dormant during the winter months. Measurements were then taken of the degree of winter growth activity for the F_1 and F_2 generations of these crosses:

	PERCENTAGE OF PLANTS ENGAGED IN DEGREE OF WINTER ACTIVITY				
	FULLY ACTIVE	FAIRLY ACTIVE	INTERMEDIATE	FAIRLY DORMANT	COMPLETELY DORMANT
P_1 (coastal)	100	0	0	0	0
P_2 (alpine)	0	0	0	0	100
F_1 ($P_1 \times P_2$)	0	0	100	0	0
F_2 ($F_1 \times F_1$)	4.4	14.3	61.1	18.3	1.4

Assuming that additive gene effects determine degree of winter activity, how many gene pairs control this trait?

7. Shown below are data for height in a hypothetical plant where highly inbred tall and short parents are crossed to produce an F_1 and the F_1 are self-fertilized to obtain an F_2.

   a. What data are needed to determine heritability of plant height and the number of loci involved in this cross?

   b. Calculate the heritability.

   c. Calculate the number of loci.

   d. Are these data appropriate to use for the variance formula?

HEIGHT (cm)	NUMBER OF PLANTS			
	P_1	P_2	F_1	F_2
20	1			
40	3			
60	7			1
80	33			2
100	84			5
120	156		2	12
140	76		5	22
160	42		11	45
180	15		37	83
200	6		86	180
220	2	1	151	202
240		7	81	179
260		11	40	92
280		40	17	45
300		82	8	20
320		149	1	10
340		83		7
360		39		3
380		12		1
400		4		
420		2		

8. Assume that variation in fruit weight in a plant species is controlled by three gene pairs and that each active (uppercase) gene acts to multiply the base weight of 4 g by 2 so that *aabbcc* = 4 g and *AABBCC* = 256 g.
   a. If *AABBCC* × *aabbcc*, give the expected F_1 weight.
   b. If the F_1 is selfed, give the expected distribution of weights in the F_2.
   c. What would be the weights of the progeny if the F_1 is:
      a) backcrossed to the light parent?
      b) backcrossed to the heavy parent?

9. The development of pulmonary tumors in mice was investigated by injecting a carcinogen and counting the tumorous nodules on the surface of the lungs 16 weeks after injection. The results for a resistant strain L, a susceptible strain A, and the F_1 and F_2 of a cross between them are shown below. Approximately 100 mice were studied in each group.

GROUP	MEAN NUMBER OF NODULES	STANDARD DEVIATION
strain L	0	0
strain A	75.4	15.7
F_1	12.5	5.3
F_2	10.0	14.1

   a. Determine the broad sense heritability of tumor resistance versus susceptibility in this cross.
   b. Determine the number of gene pairs involved.

10. The data below for ear length in corn were obtained by Emerson and East in 1913.

							LENGTH OF EARS (cm)										
	5	6	7	8	9	10	11	12	13	14	15	16	17	18	19	20	21
P_1	4	21	24	8													
P_2								3	11	12	15	26	15	10	7	2	
F_1					1	12	12	14	17	9	4						
F_2			1	10	19	26	47	73	68	68	39	25	15	9	1		

   a. Determine the mean and standard deviation of the parents, F_1, and F_2 generations.
   b. Calculate the heritability.
   c. Use the variance method to estimate the number of pairs of genes responsible for the variance seen.

11. Two homozygous inbred strains differing with respect to five gene pairs with additive effects on a quantitative trait were crossed to produce an F_1. What proportions of the F_2 (from $F_1 \times F_1$) are expected to resemble one of the homozygous inbred strains in phenotype?

12. Assume that genes A, B, C, and D have duplicate, cumulative effects and show independent assortment. Each active gene contributes 3 cm height to the organism when present. In addition, genes LL, always present in homozygous condition, contribute a constant 40 cm of height. Ignoring variation due to the environment, $AABBCCDDLL$ = 64 cm high, and $aabbccddLL$ = 40 cm high. A cross is made $AAbbCCDDLL \times aaBBccDDLL$ and carried into the F_2 generation.

    a. What height is expected in the F_1 generation?

    b. Compare the average F_1 height with the average F_2 height.

    c. What proportion of the F_2 population is expected to equal the $AAbbCCDDLL$ parent in height?

    d. What proportion of the F_2 population is expected to equal the $aaBBccDDLL$ parent in height?

13. The size of rabbits is presumably determined by genes with equal and additive effects. From a total of 2,012 F_2 progeny from crosses between large and small varieties, 6 were as small as the average small parental class, and 10 were as large as the average large parental class. How many gene pairs are causing variation in size?

14. Suppose a variety of wheat is discovered in which kernel color is determined by the action of six pairs of genes. From the cross $AABBCCDDEEFF \times aabbccddeeff$:

    a. what fraction of the F_2 is expected to resemble either parental variety?

    b. how many F_2 phenotypic classes are expected?

15. In which of the following traits is genetic variance more influential than environmental factors in determining the phenotype?

TRAIT	HERITABILITY
a. milk production in cattle	0.60
b. litter size in pigs	0.30
c. egg production in poultry	0.20
d. puberty age in rats	0.15
e. wool length in sheep	0.55
f. amount of spotting in Friesian cattle	0.95
g. body weight in sheep	0.35
h. abdominal bristle number in Drosophila	0.50

16. Plants that are $aabbddeeff$ in genotype are 9 cm tall. However, the presence of "active height genes" (represented by capital letters) adds 3 cm per gene to the base height, so a plant that is $aaBbDdeeff$ will be 15 cm tall (ignoring the effect of the environment).
    Fill in the table below, based on crosses I, II, III, and IV.

	PARENT 1		PARENT 2		OFFSPRING
CROSS I	$AaBBDdEeff$	×	$AaBBDdEeff$	→	F_1
CROSS II	$AaBbDdEeFf$	×	$AaBbDdEeFf$	→	F_1
CROSS III	$AaBbDdEeff$	×	$aabbddeeFf$	→	F_1
CROSS IV	$AAbbDDeeFF$	×	$aaBBddEEff$	→	F_1

	CROSS I	CROSS II	CROSS III	CROSS IV
height of parent 1				
height of parent 2				
height of tallest F_1				
height of shortest F_1				
proportion of F_1 that is tallest				
average height of F_1s				

Challenge Problems

17. In a certain species of plant, six pairs of genes (*AA, BB, DD, EE, GG, HH*) affect height in an equal and additive fashion. Consider the following cross:

    P: *aaBBDDeeGGHH*   ×   *AABBddeegghh*

         39 cm tall   ↓   23 cm tall

                    F_1

    a. How many centimeters does each active gene (uppercase letter) contribute to plant height?

    b. How tall is an *aabbddeegghh* plant?

    c. How tall is an *AABBDDEEGGHH* plant?

    d. How tall is an F_1 plant?

    e. What will be the ratio of *gametes* with differing numbers of active genes produced by the F_1?

    f. If an F_1 is selfed, give the expected phenotypic ratio of phenotypes in the F_2 generation.

    g. If an F_1 plant is mated to the 23 cm parental variety:

       a) give the height of the tallest offspring.

       b) give the height of the shortest offspring.

    h. If a plant heterozygous for all six gene pairs is selfed, what proportion of the progeny is expected to be 27 cm tall?

18. Stem length in a plant species can vary from 6 to 46 cm, and a polygenic system with five pairs of genes (symbolized *A, a, B, b, D, d, E, e, F, f*) is responsible for variation in stem length. Consider the following crosses:

    CROSS I   plant 1 × plant 1 → all F_1 with 30 cm stems

    CROSS II   plant 2 × plant 2 → all F_1 with 30 cm stems

    CROSS III   plant 1 × plant 2 → all F_1 with 30 cm stems

    F_1 testcross (F_1 × a plant with 6 cm stems)

                ↓

    offspring varying from 10 to 26 cm stems

    a. Give the complete genotypes for plant 1 and plant 2.

    b. In the F_1 testcross of cross III, what proportion of the offspring is expected to have 10 cm stems?

c. If the F$_1$ plants of cross III had been *self-fertilized* instead of testcrossed, give the following expected results:

    a) stem length of plants with the longest stems

    b) stem length of plants with the shortest stems

    c) proportion of offspring with 22 cm stems

    d) genotypes of the offspring with the longest stems

    e) genotypes of the offspring with the shortest stems

Team Problems

19. In a certain species of plant, seven pairs of genes (*AA, BB, DD, EE, GG, HH,* and *MM*) affect height in an equal and additive fashion. Consider the following cross:

P: *aaBBDDEEGGhhmm*   ×   *AABBddeeggHHmm*

    46 cm tall   ↓   36 cm tall

                     F$_1$

a. How tall will an *aabbddeegghhmm* plant be?

b. How tall will an *AABBDDEEGGHHMM* plant be?

c. How tall will the F$_1$ plants be?

d. If the F$_1$ plant is *self-fertilized,* calculate the phenotypes and phenotypic ratios of the expected F$_2$ plants.

e. If an F$_1$ plant is mated with the 46 cm parent:

    a) The tallest offspring would be _____cm tall.

    b) The shortest offspring would be _____cm tall.

    c) The chance of getting a 31 cm tall plant is _____.

20. Assume that three independently assorting pairs of genes act in an equal and additive fashion to affect the intensity of human skin pigmentation. Indicate in the spaces below the lightest and darkest possible skin colors among the offspring of the following groups of crosses and the greatest chance of getting the darkest possible offspring in an appropriate family. (Note: the scale of pigment intensities, from most to least, is black [with 6 active genes], very dark [with 5], dark [with 4], medium [with 3], light [with 2], very light [with 1], and white [with 0 active genes].)

GROUPS OF FAMILIES WITH PARENTS OF THE FOLLOWING SKIN COLOR INTENSITIES	LIGHTEST POSSIBLE OFFSPRING*	DARKEST POSSIBLE OFFSPRING*	CHANCE OF GETTING DARKEST POSSIBLE OFFSPRING
black × very light			
light × very light			
black × light			
black × white			
light × light			
dark × medium			
dark × white			
medium × medium			
very dark × medium			
dark × dark			

*The lightest and darkest possible offspring do not necessarily have to occur within the same family in the grouping.

Solutions

1. a. 8 cm. The F_1 genotype *AaBbCc* adds 6 cm (3 active genes × 2 cm each) to the base height of 2 cm.

    b.

FREQUENCY	NUMBER OF ACTIVE GENES	BASE HEIGHT	HEIGHT (cm)
1/64	0 (× 2 cm)	2	2
6/64	1	2	4
15/64	2	2	6
20/64	3	2	8
15/64	4	2	10
6/64	5	2	12
1/64	6	2	14

    For three heterozygous loci, $y = 3$ in Pascal's triangle, and the F_2 phenotypic ratio for height corresponds to the row with $(2y + 1) = 7$ categories.

    c. 2/64 (1/64 with 2 cm and 1/64 with 14 cm). This may also be calculated from the formula for one of the parental classes in the F_2: $(1/4)^y$, where $y =$ the number of heterozygous gene pairs. In this case, $(1/4)^3 = 1/64$. Because we are interested in *both* parental classes, the answer is $1/64 \times 2$.

    d. 0. The height of the F_1 is the result of three additive genes. It is impossible for any genotype to be homozygous for an uneven number of additive genes.

2. a. 16 cm. The F_1 from the two parental strains: *aabbcc* (2 cm) and *AABBCC* (128 cm) is *AaBbCc* (16 cm).

    b.

FREQUENCY	NUMBER OF ACTIVE GENES	BASE HEIGHT	HEIGHT (cm)*
1/64	0	2	2
6/64	1	2	4
15/64	2	2	8
20/64	3	2	16
15/64	4	2	32
6/64	5	2	64
1/64	6	2	128

    *For 2 active genes, $2 \times (2)^2$; for 3 active genes, $2 \times (2)^3$; etc.

    c. 2/64 (1/64 with 2 cm and 1/64 with 128 cm)

    d. 0 for the same reason as in problem 1.

3. a. 1/256. This comes from the binomial where:

    $$\frac{8!}{8!0!} (1/2)^8 (1/2)^0$$

    $P_r =$ probability of an active pigment gene = 1/2

    $P_s =$ probability of an inactive pigment gene = 1/2

    $n =$ total number of genes = 8 for four loci

    $r =$ number of active pigment genes for very dark = 8

    $s =$ number of inactive pigment genes for very fair = 0

b. 1/256. This comes from a binomial similar to part a except that $r = 0$ and $s = 8$.

c. 1,680/6,144 or 105/384 (27.3%). Because the intermediate parents have 4 active pigment genes, the binomial is:

$$\frac{8!}{4!4!} (1/2)^4 (1/2)^4$$

d. 336/1,536 (21.9%). The binomial is:

$$\frac{8!}{5!3!} (1/2)^5 (1/2)^3$$

4. Four gene pairs. The average number of F_2 seeds that mimic a P generation seed in weight is 4, and 1,000 F_2 seeds were observed, so:

$4/1,000 = (1/4)^y$

$1/250 = (1/4)^y$, y = approximately 4

5. Four gene pairs. The average of the most extreme F_2 classes is 7.5. $7.5/2,025 = 1/270$, which is approximately $(1/4)^y = (1/4)^4 = 1/256$.

6. Two or three gene pairs. The average percentage of F_2 that overlap one parental strain is $(4.4 + 1.4)/2 = 2.9$. So $29/1,000 = 1/34.4$, which is between $(1/4)^2 = 1/16$ and $(1/4)^3 = 1/64$.

7. a. Needed are the variances of the F_1 and F_2 to calculate broad sense heritability (H_b) and also the means of the parental strains to determine the number of loci.

b. $h^2_b = 55.1\%$, which is the ratio of the F_1 variance (918) to that of the F_2 (1,666)

c. Six or seven loci are involved: $(320 - 122)^2/[8 (1,666 - 918)] = 6.6$

d. Yes. The standard deviations of the parental strains (28.63 and 29.55) and the F_1 (30.31) are nearly identical, reflecting only environmental variation, and that of the F_2 is greater (40.82), reflecting genetic as well as environmental variation.

8. a. 32 g. The F_1 genotype $AaBbCc$ doubles the base weight of 4 g three times $(4 \times 2 \times 2 \times 2 = 32$ g).

b. 1/64 256 g: 6/64 128 g: 15/64 64 g: 20/64 32 g: 15/64 16 g: 6/64 8 g: 1/64 4 g, because $y = 3$ and $(2y + 1) = 7$ in Pascal's triangle. The heaviest F_2 has 6 multiplicative genes, each doubling the base height of 4 g $(4 \times 2 \times 2 \times 2 \times 2 \times 2 \times 2 = 256$ g), the next heaviest has 5 multiplicative genes $(4 \times 2 \times 2 \times 2 \times 2 \times 2 = 128$ g), etc.

c. a) 1/8 32 g: 3/8 16 g: 3/8 8 g: 1/8 4 g. Because one parent is $aabbcc$ (4 g), this is a testcross. The F_1 is heterozygous for three gene pairs, and the ratio of gametes with different numbers of active genes is determined from the $(2)^y = 8 = 1{:}3{:}3{:}1$ row in Pascal's triangle: 1/8 gametes have 3 multiplicative genes (32 g), 3/8 have 2 (16 g), etc.

b) 1/8 256 g: 3/8 128 g: 3/8 64 g: 1/8 32 g. The ratio of the F_1 gametes is obtained as above. In this case, the heavy parent always contributes an ABC gamete. Thus, 1/8 of the progeny have 6 multiplicative genes (256 g), 3/8 have 5 (128 g), etc.

9. a. $h^2_b = 0.86$ because:

$$h^2_b = \frac{\sigma^2_{T, F_2} - \sigma^2_{T, F_1}}{\sigma^2_{T, F_2}} = \frac{(14.1)^2 - (5.3)^2}{(14.1)^2}$$

   b. four or possibly five loci, because:
   $N = D^2/[8 \, (\sigma^2_{T, F_2} - \sigma^2_{T, F_1})]$
   $N = (75.4 - 0)^2/[8 \, (14.1)^2 - (5.3)^2]$
   $N = 5,685.16/1,365.76 = 4.2$ pairs of genes

10. a. $P_1: x = 6.6$ cm, $s = 0.8$
   $P_2: x = 16.8$ cm, $s = 1.9$
   $F_1: x = 12.1$ cm, $s = 1.5$
   $F_2: x = 12.9$ cm, $s = 2.3$

   b. $h^2_b = (2.3)^2 - (1.5)^2/(2.3)^2$
   $= 3.04/5.29 = 0.57$

   c. four or five pairs
   $N = (16.8 - 6.6)^2/8 \, [(2.3)^2 - (1.5)^2]$
   $N = 104.04/24.32 = 4.3$ pairs of genes

11. 1 in 1,024

12. a. parents: 58 cm and 52 cm, $F_1 = 55$ cm
   b. Both will be 55 cm.
   c. 15 of 64
   d. 15 of 64

13. $2,012/8 = 251$ (about 1 extreme phenotype per 256); thus, four pairs of genes were involved because $(1/4)^4 = 1/256$

14. a. 1/4,090
   b. 13

15. Heritability is the proportion of total phenotypic variance due to genetic factors. Thus, genetic influences are greater than environmental ones for traits where heritability is > 0.50. This is true for traits a, e, and f. Trait h apparently has equal genetic and environmental influences regarding phenotypic variance.

16.

	CROSS I	CROSS II	CROSS III	CROSS IV
height of parent 1	24	24	21	27
height of parent 2	24	24	12	21
height of tallest F_1	33	39	24	24
height of shortest F_1	15	9	9	24
proportion of F_1 that is tallest	1/64	1/1,024	1/32	all
average height of F_1s	24	24	16.5	24

17. a.　4 cm. The differences between the two strains are 16 cm (39 – 23) and 4 active genes (8 – 4). Therefore, 16/4 = 4 cm per active gene.

　　b.　7 cm. The 23 cm tall parent has four active genes (4 × 4 = 16 cm). So the base height is 23 – 16 = 7 cm. An *aabbddeegghh* plant is, therefore, 7 cm tall.

　　c.　55 cm, because the base height of 7 cm plus the effect of 12 additive genes (12 × 4 = 48 cm) = 55 cm

　　d.　31 cm. The genotype *AaBBDdeeGgHh* has six active genes; so 7 cm base height + (6 × 4 cm) = 31 cm.

　　e.　1/16 with five active genes: 4/16 with four: 6/16 with three: 4/16 with two: 1/16 with one. This is obtained directly from Pascal's triangle for four heterozygous loci: $y = 4$, $(2)^y = 16$, and the 1:4:6:4:1 row adds up to 16. The number of active genes is based on the four heterozygous gene pairs in the F_1 plus one *B* gene in each (because the *B* locus is homozygous, all gametes will carry *B*).

　　f.　1/256 47 cm: 8/256 43 cm: 28/256 39 cm: 56/256 35 cm: 70/256 31 cm: 56/256 27 cm: 28/256 23 cm: 8/256 19 cm: 1/256 15 cm. Because $y = 4$, $(2y + 1) = 9$ phenotypic classes, and the row in Pascal's triangle with 9 numbers is 1:8:28:56:70:56:28:8:1.

　　g.　a)　35 cm. From *AaBBDdeeGgHh* × *AABBddeegghh*, the progeny with the highest number of additive genes is *AABBDdeeGgHh* = 7 cm base height + (7 × 4 cm) = 35 cm.

　　　　b)　19 cm. From the above cross, the progeny with the lowest number of additive genes is *AaBBddeegghh* = 7 + (3 × 4 cm) = 19 cm.

　　h.　95,040/491,420 = 19.34%. For 27 cm, the base height of 7 cm must be increased by the effect of five additive genes (20 cm). The binomial is:

$$\frac{12!}{5!7!} \, (1/2)^5 \, (1/2)^7$$

18. a.　plant 1 = *AABBDDeeff*, plant 2 = *AAbbddEEFF* (plants 1 and 2 have one pair of active genes [*AA*] in common and two pairs of active genes not in common)

　　b.　1/16

　　c.　a)　46 cm

　　　　b)　14 cm

　　　　c)　28/256

　　　　d)　*AABBDDEEFF*

　　　　e)　*AAbbddeeff*

19. a. 6 cm
    b. 76 cm
    c. 41 cm
    d. F_1 is *AaBBDdEeGgHhmm*

F$_2$ PHENOTYPIC RATIOS	PHENOTYPIC HEIGHT
1/1024	16 cm
10/1024	21 cm
45/1024	26 cm
120/1024	31 cm
210/1024	36 cm
252/1024	41 cm
210/1024	46 cm
120/1024	51 cm
45/1024	56 cm
10/1024	61 cm
1/1024	66 cm

    e. a) 56 cm
       b) 31 cm
       c) 1/32

20.

GROUPS OF FAMILIES WITH PARENTS OF THE FOLLOWING SKIN COLOR INTENSITIES	LIGHTEST POSSIBLE OFFSPRING	DARKEST POSSIBLE OFFSPRING	CHANCE OF GETTING DARKEST POSSIBLE OFFSPRING
black × very light	medium	dark	1/2
light × very light	white	medium	1/8
black × light	medium	very dark	1/4
black × white	medium	medium	100%
light × light	white	dark	1/16
dark × medium	very light	black	1/32
dark × white	very light	medium	1/4
medium × medium	white	black	1/64
very dark × medium	light	black	1/16
dark × dark	light	black	1/16

11
Hardy-Weinberg Equilibrium

Problems

1. Suppose that albinism (an autosomal recessive trait, *aa*) occurs among newborns in a certain population of 10,000,000 at the frequency of 1 in 14,400. How many individuals in the population are expected to be homozygous normal (*AA*) for this trait?

2. How quickly and in what manner is genetic equilibrium reached under the following initial conditions?
    a. 1 autosomal locus, 2 alleles, different frequencies in two sexes
    b. 2 autosomal loci, 2 alleles each, not linked
    c. 1 autosomal locus, 6 alleles

3. Cystic fibrosis is a genetic disease of humans caused by homozygosity for a recessive gene (*aa*). In a particular large population, the frequency of newborns affected with cystic fibrosis is 1 in 3,600. In this population, what proportion of marriages between genetically unrelated normal individuals is at risk of producing a child affected with cystic fibrosis?

4. In some populations of grasses, the ability to grow in soil contaminated with toxic metals like nickel is a dominant trait determined by the gene *R*.
    a. In a population at genetic equilibrium, 51% of the seeds can germinate in contaminated soil. What are the frequencies of alleles *R* and *r*?
    b. What percentage of the plants that germinate in contaminated soil will be heterozygous?

5. In a certain population of garden peas, the alleles *T* for tall plants and *t* for short plants are found in frequencies of 0.70 (for *T*) and 0.30 (for *t*). If the population is in genetic equilibrium, what proportion of the progeny produced by matings of tall × tall is expected to be short?

6. The D locus controls the production of one series of antigens in the red blood cells of humans. The presence of the dominant allele *D* in the homozygous (*DD*) or heterozygous (*Dd*) condition results in the D+ phenotype, whereas the homozygous recessive (*dd*) condition results in the d– phenotype. A population was found in which 91% was D+ and 9% was d– in phenotype. Family testing also showed that of 523 matings in which one parent was D+ and the other was d–, 60% of the children were D+ in phenotype and 40% of the children were d– in phenotype.
    a. Calculate the frequencies of the *D* and *d* alleles in the population.
    b. Determine whether this population is in Hardy-Weinberg equilibrium.
    c. If the population is not in genetic equilibrium, state which genotype is in excess.

7. If a population is in Hardy-Weinberg equilibrium for the multiple alleles A^+, *A*, and *a*, whose frequencies are $p = 0.60$ for A^+, $q = 0.20$ for *A*, and $r = 0.20$ for *a*, what percentage of the population is expected to be heterozygous?

8. In a certain human population in genetic equilibrium for the typical X-linked red-green color blindness locus, 6 males in 100 are color-blind.

   a. What are the frequencies of the normal (E) and color blindness (e) alleles?

   b. What are the expected frequencies of color-blind females and heterozygous females in the population?

9. If color blindness is a recessive X-linked trait, and 18 of 20,000 females and 600 of 20,000 males in a certain population are color-blind, is the population in genetic equilibrium for the color blindness genes? Explain.

10. Pattern baldness in humans is controlled by a single autosomal locus with two alleles (B and B'), but baldness is dominant in males (BB and BB' males become bald) and recessive in females (only BB females become bald). If a population is in Hardy-Weinberg equilibrium for this locus and 4% of the women in the population become bald,

    a. what percentage of the men are bald?

    b. what percentage of the nonbald women could give a gene for baldness to their sons?

    c. what percentage of all possible marriages are expected to produce daughters who develop baldness?

11. A series of multiple alleles exists for fingernail color in martians. Three alleles are in the series: allele A1 = red nails, allele A2 = white nails, and allele A3 = blue nails. The dominance relationships of these alleles is A1 > A2 > A3. A population of martians was sampled and found to contain 360 red-, 280 white-, and 360 blue-nailed individuals.

    a. Assuming the population to be in Hardy-Weinberg equilibrium, calculate the expected frequencies of the three alleles.

    b. Give the expected frequencies of the following genotypes: A1A2, A2A3, A2A2, and A3A3.

    c. Mating data were obtained from this population: white × blue martian matings produced 400 white and 600 blue progeny. Is this population actually in genetic equilibrium for this locus? Explain.

12. In a hypothetical population, two very closely linked loci controlling hair color (dark or light) and eye color (blue or brown) were found to be in Hardy-Weinberg equilibrium when tested separately. However, when both traits were observed jointly in the population, the following data were obtained:

	BROWN HAIR	BLOND HAIR	TOTAL
brown eyes	60	10	70
blue eyes	10	20	30
total	70	30	100

    a. Is the population in linkage equilibrium for the two traits considered jointly? If not, indicate which phenotypic combinations are in excess and which are in deficiency.

    b. If they are not in linkage equilibrium, which of the following is the more probable explanation for the disparity:

       a) There are some sort of selective differences among the phenotypes.

       b) Little crossing-over occurs between the linked loci.

13. Fill in the blanks in the table below, assuming that the population is in Hardy-Weinberg equilibrium:

RECESSIVE GENETIC TRAIT	FREQUENCY OF THE RECESSIVE GENE IN THE POPULATION	FREQUENCY OF HETEROZYGOTES IN THE POPULATION	FREQUENCY OF AFFECTED INDIVIDUALS IN THE POPULATION
trait A	_____	1 in 100	_____
trait B	_____	_____	1 in 10,000
trait C	1 in 500	_____	_____
trait D	1 in 17	_____	_____
trait E	_____	1 in 125	_____

Challenge Problems

14. The following genotype frequencies were found in random samples of 100 individuals from each of four different populations. Which of these populations is not in equilibrium?

	B^1B^1	B^1B^2	B^2B^2
a.	0.3500	0.3000	0.3500
b.	0.4900	0.4200	0.0900
c.	0.0025	0.0950	0.9025
d.	0.3025	0.4950	0.2025

15. In the horse, the A locus governs the distribution of pigment in the coat. The A locus has a series of multiple alleles. These alleles control the following phenotypes: $A+$ = wild type (bay), A = dark bay (black mane and tail), a^t = seal brown, and a = recessive black. The order of dominance is $A+ > A > a^t > a$.

a. If, in a certain population, the gene frequencies are $A+ = 0.4$, $A = 0.2$, $a^t = 0.1$, and $a = 0.3$, calculate the expected frequencies of the four possible phenotypes.

b. If wild-type females from this population are captured and repeatedly mated with black males so that 800 offspring are produced, how many of the 800 are expected to be black in color? How many would be seal brown?

Team Problems

16. In humans, the drug isoniazid is used in the treatment of tuberculosis. The drug is inactivated in the liver by the enzyme acetyl transferase. Two alleles of the gene for this enzyme exist, producing enzymes that inactivate the drug rapidly (allele a^r) or slowly (allele a^s). Heterozygotes show intermediate rates of drug inactivation. In 1964, Dufor, Knight, and Harris collected the following data from three populations:

POPULATIONS	PHENOTYPES			TOTALS
	RAPID	INTERMEDIATE	SLOW	
Japanese	108	81	20	209
Caucasians	7	37	61	105
Africans	6	51	59	116

a. Which populations appear to be in genetic equilibrium?

b. Which population is most different from the others?

17. In 1967, Livingstone presented the following data for sickle-cell anemia based on samples of adult Africans and African Americans from three populations:

LOCATION	SAMPLE SIZE	NORMAL (AA)	SICKLE-CELL TRAIT (AS)	SICKLE-CELL ANEMIA (SS)
Uganda	3,362	2,817	542	3
Dallas	1,165	1,080	78	7
Philadelphia	1,000	923	74	3

Are these populations in genetic equilibrium? If not, give a plausible explanation of the disequilibrium.

18. A certain recessive trait occurs among newborn babies in the U.S. at a frequency of 1 in 4,900. Given this information and assuming that the population is in Hardy-Weinberg equilibrium:

   a. what is the frequency of the gene in question in the general population?

   b. what is the frequency of heterozygotes in the general population?

Calculate the risk that a child will display the phenotype caused by the recessive allele in question for each of the following situations:

   c. One parent is affected (has the disorder), and the other parent has no family history of the disorder.

   d. One parent has an affected sister; the other parent has no known family history of the disorder.

   e. Both parents are unrelated and have no family history of the disease.

   f. One parent is the sister of an affected individual; the other parent is a known carrier (is heterozygous).

   g. Both parents are unrelated, but each has an affected brother.

   h. One parent is heterozygous for the trait; the other parent has no known family history of the disorder.

Solutions

1. About 9,830,000. If this population is in genetic equilibrium and p = frequency of dominant allele A and q = frequency of recessive allele a, we expect genotypic frequencies of p^2 for AA, $2pq$ for Aa, and q^2 for aa. Thus, $q^2 = 1/14,400$, $q = \sqrt{1/14,400} = 1/120$.

   Because $p + q = 1.0$, $p = 1 - q$. So $p = 119/120$.

   The expected frequency of $AA = p^2 = (119/120)^2 =$ about 0.983; therefore, the expected number of AA individuals in the population is $10,000,000 \times 0.983 =$ about 9,830,000.

2.   a. Hardy-Weinberg equilibrium will be achieved in a single generation of random mating.

     b. Hardy-Weinberg equilibrium will be approached slowly, with the degree of linkage disequilibrium decreasing by the "crossover frequency" each generation (maximum of 0.50 if the loci show independent assortment). The tighter the linkage, the slower will be the approach to linkage equilibrium.

     c. Hardy-Weinberg equilibrium will be achieved in a single generation of random mating.

3. One family in 900. If we assume the population to be in genetic equilibrium, then p = frequency of normal dominant gene (A), q = frequency of cystic fibrosis recessive allele (a), and the expected genotype ratios are p^2 for AA, $2pq$ for Aa, and q^2 for aa. So $q^2 = 1/3600$, $q = \sqrt{1/3600} = 1/60$.

   Because $p + q = 1.0$, $p = 1 - q$. So $p = 59/60$, $2pq$ = approximately $2q$ (because p is very nearly 1), $2q = 2/60 = 1/30$, the expected frequency of heterozygotes (Aa) in the population.

   Because marriages between genetically unrelated normal individuals are at risk of producing children with cystic fibrosis only if both parents are heterozygous, the risk of such marriages is $Aa \times Aa = 1/30 \times 1/30 = 1$ family in 900.

4. If 51% of the seeds are resistant, 49% are sensitive. Thus:

   a. Resistant seeds = p^2 (for RR) + $2pq$ (for Rr) = 0.51.

      Sensitive seeds = q^2 (for rr) = 0.49.

      q = frequency of allele r = $\sqrt{0.49}$ = 0.70.

      p = frequency of allele R = $1 - 0.70$ = 0.30.

   b. 82.4%. Of the plants that grow in contaminated soil, heterozygotes (Rr) = $2pq = 2 \times 0.30 \times 0.70$ = 0.42, and homozygotes (RR) = $p^2 = (0.3)^2$ = 0.09. Thus, the percentage of heterozygous resistant plants is 0.42/0.51 = 0.824 or 82.4%.

5. About 5.3%. Because $p = 0.70$ (for T) and $q = 0.30$ (for t), we expect the genotype frequencies to be p^2 for TT, $2pq$ for Tt, and q^2 for tt. Four types of tall $\times$ tall matings will take place at the following expected frequencies:

MALE PARENT		FEMALE PARENT	TOTAL FREQUENCIES	FREQUENCY OF SHORT OFFSPRING
TT	$\times$	TT	$0.49 \times 0.49 = 0.2401$	0.0
TT	$\times$	Tt	$0.49 \times 0.42 = 0.2058$	0.0
Tt	$\times$	TT	$0.42 \times 0.49 = 0.2058$	0.0
Tt	$\times$	Tt	$0.42 \times 0.42 = \underline{0.1764}$	$\underline{0.0441}$
		totals:	0.8281	0.0441

   Assuming all matings produce equal numbers of offspring, the proportion of short offspring expected from these matings will be 0.0441/0.8281 = 0.0533 or about 5.3%.

6. a. Assuming genetic equilibrium, q^2 = frequency of dd = 0.09.

      So q = frequency of allele d = $\sqrt{0.09}$ = 0.30, and because $p + q = 1.00$, p = frequency of allele D = $1 - 0.30 = 0.70$.

   b. Expected frequencies of $DD = p^2 = 0.49$ and $Dd = 2pq = 0.42$. The expected ratio of DD to Dd is 0.49/0.91 = 7/13 for DD to 0.42/0.91 = 6/13 for Dd, or 7:6. Considering the mating data:

PARENTS	EXPECTED RATIO OF MATINGS	EXPECTED FREQUENCIES OF OFFSPRING	
		$D-$	dd
$DD \times dd$	7/13	7/13	0
$Dd \times dd$	6/13	$\underline{3/13}$	$\underline{3/13}$
	expected totals	10/13 = 0.769	3/13 = 0.231
	observed totals	0.600	0.400

   Comparison of the observed and expected frequencies of offspring shows considerable disagreement between them. Therefore, it appears that the population is not in genetic equilibrium.

   c. Genotype dd individuals appear to be in excess in this population.

7. In this population, where $p = 0.60$ for A^+, $q = 0.20$ for A, and $r = 0.20$ for a, we expect $(p + q + r)^2$ $= p^2$ for A^+A^+ + q^2 for AA + r^2 for aa + $2pq$ for A^+A + $2pr$ for A^+a + $2qr$ for Aa = 1.00. The frequency of heterozygotes, therefore, is 0.24 for A^+A + 0.24 for A^+a + 0.08 for Aa = 0.56 or 56%.

8. Because the locus for red-green color blindness is X-linked, males are hemizygous. Thus, p = frequency of normal (DY) males and q = frequency of color-blind (dY) males. However, among the females, p^2 = frequency of DD, $2pq$ = frequency of Dd, and q^2 = frequency of dd females. So in this population:

   a. Because 6 of 100 males are color-blind, q = frequency of allele d = 6/100 = 0.06, and p = frequency of allele D = 1 – 0.06 = 0.94.

   b. Among the females, we expect p^2 for homozygous normal (DD) = 0.884 or 88.4%, $2pq$ for heterozygous normal (Dd) = 0.113 or 11.3%, q^2 for color-blind (dd) = 0.0036 or about 0.4%.

9. Frequency of color-blind (ee) females = p^2, p^2 = 18/20,000 = 0.0009, so p in females = $\sqrt{0.0009}$ = 0.03. Frequency of color-blind (eY) males = p, p = 600/20,000 = 0.03. Because p in females and males is the same, the population is in genetic equilibrium.

10. In this population, we expect the following situation for pattern baldness:

| | **GENOTYPE** | | |
	BB	BB'	$B'B'$
frequencies	p^2	$2pq$	q^2
males	bald	bald	nonbald
females	bald	nonbald	nonbald

Because 4% of the women are bald, p^2 = frequency of bald women = 0.04, p = frequency of allele $B = \sqrt{0.04} = 0.20$, q = frequency of allele B' = 1.0 – 0.2 = 0.80.

   a. Percentage of bald men = p^2 (for BB) + $2pq$ (for BB') = 0.04 + 0.32 = 0.36 or 36%.

   b. 33.3% = 1/3. Only heterozygous (BB') nonbald women could give allele B to their sons, causing their baldness. Then the frequency of BB' females = $2pq$ = 0.32 or 32% of all females. Counting only nonbald females, however, we have $2pq$ (for BB') + q^2 (for $B'B'$) = 0.32 + 0.64 = 0.96. So 1/3 (0.32/0.96) of nonbald women can give a baldness gene to their sons.

   c. 12.96%. Only the following marriages can produce daughters who develop baldness (BB):

FATHERS		MOTHERS	FREQUENCIES		TOTALS
BB (p^2)	×	BB' ($2pq$)	0.04 × 0.32	=	0.0128
BB (p^2)	×	BB (p^2)	0.04 × 0.04	=	0.0016
BB' ($2pq$)	×	BB' ($2pq$)	0.32 × 0.32	=	0.1024
BB' ($2pq$)	×	BB (p^2)	0.32 × 0.04	=	0.0128
				total	0.1296

11. a.   Let p = frequency of allele $A1$, q = frequency of allele $A2$, and r = frequency of allele $A3$. Using the trinomial $(p + q + r)^2$, we get the following:

| | **PHENOTYPES** | | |
	RED	**WHITE**	**BLUE**
genotypes and expected frequencies	$A1A1 = p^2$ $A1A2 = 2pq$ $A1A3 = 2pr$	$A2A2 = q^2$ $A2A3 = 2qr$	$A3A3 = r^2$
observed numbers	360	280	360
observed frequencies	0.360	0.280	0.360

Frequency of allele $A3 = r = \sqrt{r^2} = \sqrt{0.360} = 0.60$. Frequency of allele $A2$: white + blue = $q^2 + 2qr + r^2 = (q + r)^2 = 0.280 + 0.360$, $(q + r)^2 = 0.640$, $(q + r) = \sqrt{0.640} = 0.80$, $q + r = 0.80$, $q = 0.80 - r = 0.80 - 0.60 = 0.20$.

Frequency of allele $A1$: $p + q + r = 1.00$, $p = 1.00 - q - r = 1.00 - 0.20 - 0.60$, $p = 0.20$.

  b.   Frequency of $A1A2 = 2pq = 2 (0.20) (0.20) = 0.08$. Frequency of $A2A3 = 2qr = 2 (0.20) (0.60) = 0.24$. Frequency of $A2A2 = q^2 = (0.20)^2 = 0.04$. Frequency of $A3A3 = r^2 = (0.60)^2 = 0.36$.

  c.   Expected frequencies of $A2A2 = 0.04$ and $A2A3 = 0.24$; relative ratios of $A2A2$ to $A2A3 = 0.04/0.28$ to $0.24/0.28 = 1/7$ to $6/7$, or 1:6.

From the mating data:

| PARENTS | EXPECTED RATIO OF MATINGS | EXPECTED FREQUENCIES OF OFFSPRING | |
		WHITE	**BLUE**
$A2A2 \times A3A3$	1/7	1/7	0
$A2A3 \times A3A3$	6/7	3/7	3/7
	expected frequencies	4/7 = 0.571	3/7 = 0.429
	expected numbers	571	429
	observed numbers	400	600

Chi-square analysis:

PHENOTYPES	OBSERVED	EXPECTED	$(o - e)$	$(o - e)^2$	$(o - e)^2/E$
white	400	571	−171	29,241	51.21
blue	600	429	171	29,241	68.32
totals	1,000	1,000			$chi^2 = 119.53$

Note: degrees of freedom = 1, P <<< 0.01.

Thus, the population is not in Hardy-Weinberg equilibrium because the expected and observed numbers of types of offspring are in significant disagreement.

12. a.   The population is not in genetic equilibrium for, as the following analysis indicates, there is an excess of brown-brown and blue-blond combinations and a deficiency of the other types:

| PHENOTYPES | | OBSERVED | EXPECTED |
HAIR	**EYES**	**NUMBERS**	**NUMBERS***
brown	brown	60	$(70 \times 70)/100 = 49$
brown	blue	10	$(30 \times 70)/100 = 21$
blond	brown	10	$(70 \times 30)/100 = 21$
blond	blue	20	$(30 \times 30)/100 = 9$

*Total number of brown hair × total number of brown eyes divided by total number in the sample = expected number with brown hair and brown eyes. Others are calculated in a similar manner.

b. b is not likely because, given enough time, linkage equilibrium will occur despite the tightness of any potential linkage. a is more likely because perhaps blue-blond and brown-brown are more attractive combinations of traits for potential mates than the other types. This sort of nonrandom mating could lead to the observed linkage disequilibrium.

13.

RECESSIVE GENETIC TRAIT	FREQUENCY OF THE RECESSIVE GENE IN THE POPULATION = q	FREQUENCY OF HETEROZYGOTES IN THE POPULATION = $2q$	FREQUENCY OF AFFECTED INDIVIDUALS IN THE POPULATION = q^2
trait A	1 in 200	1 in 100	1 in 40,000
trait B	1 in 100	1 in 50	1 in 10,000
trait C	1 in 500	1 in 250	1 in 250,000
trait D	1 in 17	32 in 189*	1 in 289
trait E	1 in 250	1 in 125	1 in 62,500

*However, with trait D, $q = 1/17$ (about 7%) and $p = 16/17$ (about 93%). Because p is not nearly 1 in this case, we must use $2pq$.

The frequency of the recessive gene in the population is q. The frequency of heterozygotes in the population is approximately $2q$ (when p is very close to 1.0 and q is very close to 0). The frequency of affected individuals in the population is q^2.

14. In each case, let p = frequency of B^1 = frequency of B^1B^1 + 1/2 frequency of B^1B^2, and q = frequency of B^2 = frequency of B^2B^2, + 1/2 frequency of B^1B^2. So

a. Observed $p = 0.35 + 0.15 = 0.50$, observed $q = 0.35 + 0.15 = 0.50$, expected $B^1B^1 = p^2 = 0.25$, expected $B^1B^2 = 2pq = 0.50$, expected $B^2B^2 = q^2 = 0.25$. Thus, the population is not in Hardy-Weinberg equilibrium because observed and expected phenotype frequencies are not the same.

b. Observed $p = 0.70$ and observed $q = 0.30$, expected $B^1B^2 = 0.49$, $B^1B^2 = 0.42$, and $B^2B^2 = 0.09$. Thus, the population is in equilibrium.

c. Observed $p = 0.05$ and observed $q = 0.95$, expected $B^1B^1 = 0.0025$, $B^1B^2 = 0.0950$, and $B^2B^2 = 0.9025$. Thus, the population is in equilibrium.

d. Observed $p = 0.55$ and observed $q = 0.45$, expected $B^1B^1 = 0.3025$, $B^1B^2 = 0.4950$, and $B^2B^2 = 0.2025$. Thus, the population is in equilibrium.

15. Let frequency of allele $A^+ = p = 0.4$; frequency of allele $A = q = 0.2$; frequency of allele $a^t = r = 0.1$; frequency of allele $a = s = 0.3$; $(p + q + r + s)^2 = p^2$ for A^+A^+, q^2 for AA, r^2 for a^ta^t, s^2 for aa, $2pq$ for A^+A, $2pr$ for A^+a^t, $2ps$ for A^+a, $2qr$ for Aa^t, $2qs$ for Aa, and $2rs$ for a^ta.

a.

PHENOTYPES	EXPECTED FREQUENCIES AT EQUILIBRIUM
wild type	A^+A^+ + A^+A + A^+a^t + A^+a p^2 + $2pq$ + $2pr$ + $2ps$ 0.16 + 0.16 + 0.08 + 0.24 = 0.64
dark bay	AA + Aa^t + Aa q^2 + $2qr$ + $2qs$ 0.04 + 0.04 + 0.12 = 0.20
seal brown	a^ta^t + a^ta r^2 + $2rs$ 0.01 + 0.06 = 0.07
black	aa = s^2 = 0.09

b.

PARENTS	EXPECTED RATIO OF MATINGS	EXPECTED FREQUENCIES OF OFFSPRING			
		BLACK	SEAL BROWN	WILD	BAY
$A^+A^+ \times aa$	0.16/0.64 = 2/8	0	0	4/16	0
$A^+A \times aa$	0.16/0.64 = 2/8	0	0	2/16	2/16
$A^+a \times aa$	0.24/0.64 = 3/8	3/16	0	3/16	0
$A^+a^t \times aa$	0.08/0.64 = 1/8	0	1/16	1/16	0
		3/16	1/16	10/16	2/16
total expected of 800 offspring		150	50	500	100

16. a.

	JAPANESE	CAUCASIANS	AFRICANS
q = frequency of s ($ss + 1/2\ rs$)	0.289	0.757	0.728
p = frequency of r ($rr + 1/2\ rs$)	0.711	0.243	0.272
p^2 = expected frequency of rr	0.51	0.06	0.07
expected numbers of rr	106.6	6.3	8.1
$2pq$ = expected frequency of rs	0.41	0.37	0.40
expected numbers of rs	85.7	38.9	46.4
q^2 = expected frequency of ss	0.08	0.57	0.53
expected numbers of ss	16.7	59.8	61.5

Chi-square analysis of Japanese:

GENOTYPES	OBSERVED	EXPECTED	$(o - e)$	$(o - e)^2$	$(o - e)^2/E$
rr	108	106.6	1.4	1.96	0.02
rs	81	85.7	–4.7	22.09	0.26
ss	20	16.7	3.3	10.89	0.65
	209	209.0			chi^2 = 0.93

Note: degrees of freedom = 1, 0.50 > P > 0.30.

Chi-square analysis of Caucasians:

GENOTYPES	OBSERVED	EXPECTED	$(o - e)$	$(o - e)^2$	$(o - e)^2/E$
rr	7	6.3	0.7	0.49	0.08
rs	37	38.9	–1.9	3.61	0.09
ss	61	59.8	1.2	1.44	0.02
	105	105.0			chi^2 = 0.19

Note: degrees of freedom = 1, 0.70 > P > 0.50.

Chi-square analysis of Africans:

GENOTYPES	OBSERVED	EXPECTED	$(o - e)$	$(o - e)^2$	$(o - e)^2/E$
rr	6	8.1	–2.1	4.41	0.54
rs	51	46.4	4.6	21.16	0.46
ss	59	61.5	–2.5	6.25	0.10
	116	116.0			chi^2 = 1.10

Note: degrees of freedom = 1, 0.30 > P > 0.20.

Thus, all three populations are in Hardy-Weinberg equilibrium.

b. Although all three populations are in genetic equilibrium, the Japanese population is most different from the others in frequencies of the alleles and genotype frequencies.

17.

	UGANDA	DALLAS	PHILADELPHIA
q = frequency of S ($SS + 1/2\ AS$)	0.0815	0.0395	0.0400
p = frequency of A ($AA + 1/2\ AS$)	0.9185	0.9605	0.9600
p^2 = expected frequency of AA	0.8436	0.9226	0.9216
expected numbers of AA	2,836.18	1,074.83	921.60
$2pq$ = expected frequency of AS	0.1497	0.0759	0.0768
expected numbers of AS	503.29	88.42	76.80
q^2 = expected frequency of SS	0.0066	0.0015	0.0016
expected numbers of SS	22.33	1.75	1.60

Chi-square analysis of Uganda:

GENOTYPES	OBSERVED	EXPECTED	$(o - e)$	$(o - e)^2$	$(o - e)^2/E$
AA	2,817	2,836.18	−19.18	367.87	0.13
AS	542	503.29	38.71	1,498.46	2.98
SS	3	22.33	−19.33	373.65	16.73
	3,362	3,361.80			chi^2 = 19.84

Note: degrees of freedom = 1, P << 0.001.

Chi-square analysis of Dallas:

GENOTYPES	OBSERVED	EXPECTED	$(o - e)$	$(o - e)^2$	$(o - e)^2/E$
AA	1,080	1,074.83	5.17	26.73	0.02
AS	78	88.42	−10.42	108.58	1.23
SS	7	1.75	5.25	27.56	15.75
	1,165	1,165.00			chi^2 = 17.00

Note: degrees of freedom = 1, P << 0.001.

Chi-square analysis of Philadelphia:

GENOTYPES	OBSERVED	EXPECTED	$(o - e)$	$(o - e)^2$	$(o - e)^2/E$
AA	923	921.6	1.4	1.96	0.002
AS	74	76.8	−2.8	7.84	0.102
SS	3	1.6	1.4	1.96	1.225
	1,000	1,000.0			chi^2 = 1.329

Note: degrees of freedom = 1, 0.30 < P < 0.20.

Thus, the Uganda population is not in Hardy-Weinberg equilibrium. There is a significant excess of heterozygotes and deficiency of homozygotes with sickle-cell anemia. This may be due to natural selection. The Dallas population is not in Hardy-Weinberg equilibrium. There is a significant deficiency of homozygotes with sickle-cell anemia. This may be due to natural selection. The Philadelphia population is in Hardy-Weinberg equilibrium.

18. a. 1 in 70 = $\sqrt{1/4,900}$ = q

b. 1 in 35 = $2q$

c. 1 in 70 = $1/35 \times 1/2$

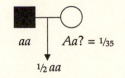

aa $Aa?$ = $1/35$

$1/2\ aa$

d. 1 in 210 = 1/35 × 2/3 × 1/4

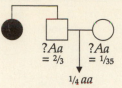

$?Aa$
$= 2/3$ $?Aa$
            $= 1/35$

$1/4\, aa$

e. 1 in 4,900 = 1/35 × 1/35 × 1/4

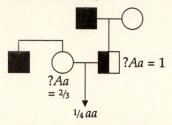

$?Aa$
$= 1/35$ $?Aa$
            $= 1/35$

$1/4\, aa$

f. 1 in 6 = 2/3 × 1/4

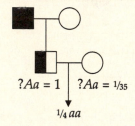

$?Aa = 1$

$?Aa$
$= 2/3$

$1/4\, aa$

g. 1 in 9 = 2/3 × 2/3 × 1/4

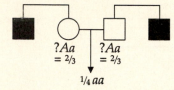

$?Aa$
$= 2/3$ $?Aa$
            $= 2/3$

$1/4\, aa$

h. 1 in 140 = 1/35 × 1/4

$?Aa = 1$ $?Aa = 1/35$

$1/4\, aa$

Evolution

Problems

1. **MATCHING TEST**   Choose the single best answer from the choices to the right. Choices may be used more than once.

   ___ will lead to the elimination of one allele from a population, but one cannot predict which one will be eliminated

   ___ is called "disruptive selection"

   ___ will eliminate an allele in a single generation

   ___ will produce a stable equilibrium with both alleles still present

   ___ will reduce the frequency of one allele in a population, but very slowly

   ___ is called "heterozygous advantage"

   a.   selection against homozygotes
   b.   selection against heterozygotes
   c.   selection against dominants
   d.   selection against recessives

2. **MATCHING TEST**

   ___ causes change in gene frequency predictable in amount, but not direction

   ___ introduces new variation from extrapopulational sources

   ___ introduces new variation from intrapopulational sources

   ___ can lead to "nonadaptive" evolution

   ___ is effective only in large populations

   ___ retards divergence among different populations of the same species

   ___ the probable reason that small, isolated American Indian tribes differ widely in blood group frequencies

   a.   migration
   b.   mutation
   c.   genetic drift
   d.   natural selection
   e.   none of these

3. **MATCHING TEST**

  __ populations will eventually become homozygous for allele A

  __ populations will eventually becom homozygous for allele a

  __ a population will reach stable equilibrium only when it becomes homozygous

  __ a population will be in genetic equilibrium only when the frequency of allele $A = 0$

  __ a population will reach stable equilibrium even though both alleles A and a are still present in the population

a. $A \underset{v}{\overset{u}{\longleftrightarrow}} a$

b. $A \overset{u}{\longrightarrow} a$ only

c. $a \overset{v}{\longrightarrow} A$ only

d.   both b and c

e.   none of these

4. **MATCHING TEST**

  __ stabilizing selection

  __ directional selection

  __ disruptive selection

  __ stable equilibrium

  __ unstable equilibrium

  __ selection versus dominants

  __ selection versus recessives

  __ heterozygous advantage

  __ allele a will be reduced by selection

  __ allele A will be eliminated by selection

  __ either A or a will be eliminated by selection, but the outcome is uncertain initially

  __ neither A nor a will be eliminated by selection

  __ both A and a will be eliminated by selection in the same population

FITNESSES

	AA	Aa	aa
a.	0.0	1.0	0.0
b.	1.0	1.0	0.0
c.	0.0	0.0	1.0
d.	1.0	0.5	1.0
e.	both b and c		
f.	none of these		

5. If the mutation rate from A to a is 10^{-5} and from a to A is 10^{-3}, what will be the frequency of allele A in a population that has reached genetic equilibrium?

6. If the mutation rate of the normal allele A to the recessive allele a is 7×10^{-6} and the reverse mutation rate of a to A is 1×10^{-7}, what will be the equilibrium frequency of gene a due to mutation in a randomly breeding population not influenced by natural selection?

7. Ninety rats live on an island, and the frequency of a recessive coat color allele in this population is 0.80. Ten rats from the mainland are inadvertently brought to the island by a ferry boat. The frequency of the coat color allele in the mainland rat population is 0.20. After the migration, what is the frequency of the coat color allele in the new rat population on the island?

8. If the frequency of blood type allele *B* is 0.3 in a pure white population and 0.8 in a pure black population and 0.6 in a mixed black population, what average percentage of the *B* alleles in the mixed black population came from white ancestry?

9. If the fitnesses of three types of flies are 0.8 for *AA*, 1.0 for *Aa*, and 0.4 for *aa* and we start with a population in which the frequency of allele *A* is 0.3 and the frequency of allele *a* is 0.7, what will be the expected frequency of allele *A* in the next generation?

10. When the population in Problem 9 reaches genetic equilibrium, what will be the frequency of allele *A*?

11. An acute viral disease infects a population of chipmunks, and only those homozygous for allele *b* are resistant. (Those of genotypes *Bb* and *BB* are susceptible and suffer 100% mortality at birth.) If the initial frequency of allele b is 0.20, what is the frequency (ignoring mutation) one generation after the introduction of this disease?

12. In a South American species of fruit fly, body color is controlled by a single pair of alleles, the body color being dark (*EE*), intermediate (*Ee*), or light (*ee*). Researchers found that a lab population of offspring from matings between intermediate body color flies (*Ee* × *Ee*) consisted of 3,100 dark, 3,500 intermediate, and 3,400 light-colored flies.
    a. Determine the fitness and selection coefficient values for all three phenotypes, and indicate which sort of selection is in operation.
    b. What will happen to this population relative to the *E* and *e* alleles if it is allowed to continue existing in the laboratory?

13. In a South American species of fruit fly closely related to the one in Problem 12, the same body color polymorphism was examined. A lab population of offspring from *Ee* × *Ee* matings consisted of 1,605 dark, 3,767 intermediate, and 1,310 light-colored flies.
    a. Determine the fitness and selection coefficient values for all three phenotypes, and indicate which sort of selection is in operation.
    b. What will happen to this population relative to the *E* and *e* alleles if it is allowed to continue breeding in the laboratory?
    c. If a genetic equilibrium is expected, give the equilibrium values for any alleles that will persist.

14. Black wool is produced in sheep that are homozygous for a recessive gene. Suppose you bought a large flock of sheep in which 25% of the animals had black wool.
    a. How many generations of complete selection would be necessary before the frequency of sheep with black wool is reduced to 4%?
    b. How many further generations would it then take to reduce the frequency of black sheep to 1%?

15. In a certain population of rats, a recessive allele *s* causes sterility in both males and females when homozygous. The frequency of gene *s* in a certain population is 0.01.
    a. Determine the mutation rate of gene *S* to gene *s* in this population.
    b. Suppose the mutation rate increased 10-fold in this population. Determine the new equilibrium frequency for gene *s* and the frequency of sterile individuals at equilibrium.

16. Generally speaking, cystic fibrosis (CF, an autosomal recessive condition) occurs in American white populations at the rate of 1 out of every 3,600 newborn infants. If we assume that CF homozygotes do not reproduce and that the condition has no effect on heterozygotes, what will be the expected rate of infants born with CF 50 generations from now?

Challenge Problem

17. How many generations will it take to reduce the frequency of births of individuals homozygous for a certain recessive trait from 1 in 40,000 to 1 in 160,000 if all individuals homozygous for this trait are sterilized at birth each generation?

Team Problems

18. **TRUE-FALSE**   In each case, indicate whether the statement is true or false. If false, explain why.

   a. If an allele is present in a population at very low frequency, there will be more homozygotes than heterozygotes for that allele, but the situation will be reversed if the allele is present at a very high frequency.

   b. Adaptations may be validly considered independent of the environments in which the organisms live.

   c. Disruptive selection in a two-allele system may result in the elimination of one allele, but it is impossible to predict which allele will be eliminated.

   d. Selection favoring heterozygotes but not homozygotes in a population will lead to the maintenance of both alleles in an equilibrium condition.

   e. If genotype AA females leave, on the average, 20 offspring and aa females leave a mean of 25 offspring, then AA has a selection coefficient of 0.8, and aa has a fitness of 1.0.

   f. Suppose eugenic selection is practiced only against homozygous individuals affected by a certain recessive allele. If the frequency of individuals showing the homozygous recessive phenotype is 1 in 40,000, it will take 50,000 generations to reduce the frequency of such individuals to 1 in 90,000.

   g. If the homozygous recessive phenotype is at a selective disadvantage, the recessive allele will be eliminated at a faster rate than a dominant allele will be eliminated if the dominant phenotype were at a severe disadvantage.

   h. It is hard to conceive of a situation in which eugenic selection would be effective in quickly eliminating all previously existing deleterious genes from a population.

   i. Population 1 has a frequency for allele A of 0.8. Population 2 has a frequency of 0.3 for allele A. The rate of migration of individuals from population 2 into 1 is 0.2. After one generation of migration, the frequency of allele A in population 1 will be 0.55.

   j. If the present frequency of an allele in a "mixed" black population is known, as is the frequency of that allele in the "pure" black ancestral population, one has sufficient information to determine the amount of gene flow into the "mixed" population.

   k. Genetic drift is called the "dispersive factor" because in a single small population, there will be an increase in variability predictable in amount but not in direction due to genetic drift.

   l. As the size of a population increases, the changes in gene frequencies due to nongenetic-caused death will tend to increase.

   m. In a modern explanation of evolution, the sources of new gene variation are mutation, independent assortment, and crossing-over.

   n. The major reason that most deleterious recessive alleles are at a low frequency in most populations is the very low mutation rates that produce them.

o. The initial allele frequencies in a population will determine the equilibrium allele frequencies in that population if bidirectional recurrent mutation occurs.

p. Unidirectional recurrent mutation will cause a change in gene frequency, but eventually a stable equilibrium will be established so that both alleles will be maintained in the population.

19. Hurler syndrome is an autosomal recessive disease in which mucopolysaccharides accumulate in body tissues, leading to moderate dwarfism, grotesque facial appearance, joint contractures, mental retardation, and early death. Suppose that the mutation rate of gene H (normal) to gene h (Hurler syndrome) is about 0.9×10^{-5}.

a. No effective medical treatment is currently available for this condition. What is the expected equilibrium frequency of gene h? How many cases of Hurler syndrome will occur each generation in a population of 10 million?

b. Suppose a medical treatment is developed that allows children with Hurler syndrome to survive and reproduce so that the fitness of these individuals increases to 0.5. In this situation, what is the expected equilibrium frequency of gene h, and how many cases will occur each generation in a population of 10 million?

c. How many generations will it take to reach the new equilibrium frequency (part b) once the new treatment is developed?

Solutions

1. b, b, c, a, d, a

2. c, a, b, c, e, a, c

3. c, b, d, b, a

4. a, e, d, a, d, c, b, a, b, c, d, a, f

5. The frequency of A will be 0.99. At equilibrium,

$$\hat{p} = v/(u + v)$$
$$= 100 \times 10^{-5}/(1 \times 10^{-5} + 100 \times 10^{-5})$$
$$= 100/101 = 0.99$$

$$\hat{q} = u/(u + v) = 1/101 = 0.01$$

6. 0.986, because

$$\hat{q} = u/(u + v)$$
$$= 7 \times 10^{-6}/(7 \times 10^{-6} + 0.1 \times 10^{-6})$$
$$= 7/7.1 = 0.986$$

7. 0.74

$$M \text{ (rate of migration)} = 10/(90 + 10) = 0.10$$

$$q \text{ (original frequency on the island)} = 0.80$$

$$Q \text{ (mainland frequency)} = 0.20$$

$$q_1 = M(Q - q_0) + q_0$$

$$= (0.10)(0.20 - 0.80) + 0.80$$

$$q_1 = 0.74$$

The initial *difference* in frequency of the recessive allele between the populations before migration was 0.60 (0.80 – 0.20), but after migration, it is 0.54 (0.74 – 0.20), a decrease of 0.06. As illustrated, one effect of migration is to decrease genetic differences between the populations involved.

8. 40

$$M = (q_b - q)/(Q - q)$$

$$= (0.6 - 0.8)/(0.3 - 0.8)$$

$$= 0.40 = 40\%$$

9. 0.41. For heterozygous advantage,

$$\Delta q = pq(ps_1 - qs_2)/1 - p^2 s_1 - q^2 s_2$$

$$= \frac{(.3)(.7)[(.3)(.2) - (.7)(.6)]}{1 - (.3)^2(.2) - (.7)^2(.6)} = -0.11$$

so,   new p = the frequency of allele a

$$= 0.70 - 0.11 = 0.59$$

new q = the frequency of allele A

$$= 1 - 0.59 = 0.41$$

10. At equilibrium,

$$\hat{q} = s_1/(s_1 + s_2) = 0.2/(0.2 + 0.6) = 0.25$$

$$p = s_2/(s_1 + s_2) = 0.6/(0.2 + 0.6) = 0.75$$

11. Allele b will increase in frequency to 100% because all chipmunks carrying the B allele suffer 100% mortality, assuming that mortality occurs before reproduction is possible.

12. a. In this case, we expect 0.25 *EE*, 0.50 *Ee*, and 0.25 *ee* offspring from the *Ee* × *Ee* matings.

	GENOTYPES		
	EE	*Ee*	*ee*
observed frequencies	.310	.350	.340
expected frequencies	.250	.500	.250
survival rates	1.240	.700	1.360
fitnesses	.912	.515	1.000
selection coefficients	.088	.485	0.000

Disruptive selection is taking place because heterozygotes have the lowest fitness.

   b. Theoretically, the population will split into two separate breeding units, one consisting of *EE* individuals and the other consisting of *ee* individuals.

13. a.

	GENOTYPES		
	EE	*Ee*	*ee*
observed frequencies	.240	.564	.196
expected frequencies	.250	.500	.250
survival rates	.960	1.128	.784
fitnesses	.851	1.000	.695
selection coefficients	.149	0.000	.305

In this case, heterozygous advantage is taking place.

b. This population is expected to establish a stable equilibrium with both the E and e alleles remaining.

c. At genetic equilibrium, let q = the frequency of allele e and p = the frequency of allele E:

$\hat{q} = s_1/(s_1 + s_2) = 0.305/(0.305 + 0.149) = 0.672$

$\hat{p} = s_2/(s_1 + s_2) = 0.149/(0.305 + 0.149) = 0.328$

14. If 25% of the population has black wool and the population is in Hardy-Weinberg equilibrium,

q^2 = frequency of black sheep = 0.25

q = frequency of recessive allele = 0.50 = 1/2

a. After selection,

$q^2 = 0.04$, $q = 0.20 = 1/5$

$t = \dfrac{1}{q_{t2}} - \dfrac{1}{q_{t1}} = 5 - 2 = 3$ generations

b. After further selection,

$q^2 = 0.01$, $q = 0.10 = 1/10$

$t = \dfrac{1}{q_{t2}} - \dfrac{1}{q_{t1}} = 10 - 5 = 5$ generations

15. a. 1×10^{-4}. Assuming the population to be in Hardy-Weinberg equilibrium, if the frequency of s (= q) is 0.01, the frequency of sterile rats ($ss = q^2$) is 0.0001 (1 in 10,000). Because $\hat{q} = \sqrt{u/s}$ and $s = 1$ (*aa* rats are sterile), $0.01 = \sqrt{u/1}$, $u = (0.01)^2 = 0.0001 = 1 \times 10^{-4}$.

b. If u increases 10-fold (from 0.0001 to 0.001),

$\hat{q} = \sqrt{u/s} = \sqrt{0.001/1} = 0.032$

So the new equilibrium frequency of gene s is 0.032, and the expected frequency of sterile rats is $q^2 = 0.001 = 1$ in 1,000.

16. 1 in 12,100

Use the formula: $t = \dfrac{1}{q_{t2}} - \dfrac{1}{q_{t1}}$

If $q_{t2} = (1/x)$ and solving for x,

$50 = \dfrac{1}{(1/x)} - \dfrac{1}{(1/60)}$

$50 = x - 60$

$x = 50 + 60 = 110$

$q_{t2} = 1/110 = 0.009$

So 50 generations from now, affected infants (*aa*) will be born at the rate of $q^2 = (1/110)^2 = 1$ in 12,100.

17. 200 generations

$$t = (1/q_t) - (1/q_0)$$
$$= [1/(1/400)] - [1/(1/200)]$$
$$= 400 - 200 = 200 \text{ generations}$$

18. a. **False,** because heterozygotes ($2pq$) are expected to outnumber homozygotes (p^2 or q^2) for an allele at low frequency, but the reverse is true if the allele is at high frequency

b. **False,** because the adaptiveness of a trait depends on the environmental situation as well as its immediate phenotypic effect

c. **True**

d. **True**

e. **False,** because fitness of AA is $20/25 = 0.80$, and the selection coefficient for AA will be 0.20

f. **False,** because

$$t = (1/q_t) - (1/q_0) = [1/(1/300)] - [1/(1/200)]$$
$$= 300 - 200$$
$$= 100 \text{ generations}$$

g. **False,** because dominant alleles are eliminated by selection faster than recessive alleles under severe selection pressure

h. **False,** because eugenic selection against a dominant phenotype would eliminate preexisting dominant alleles very quickly

i. **False,** because

$$\Delta q = M (Q - q)$$
$$= 0.2 (0.3 - 0.8)$$
$$\Delta q = -0.10$$

So gene A will decrease in frequency from 0.80 to 0.70.

j. **False,** because one also needs to know the frequency of the allele in the present white population

k. **False,** because eventually all gene loci will become homozygous for one of the allele types originally present, although which allele will persist cannot be predicted with certainty beforehand

l. **False,** because the effect of genetic drift increases with decreasing population size

m. **False,** because only mutation produces "new" allele types from within the population; independent assortment and crossing-over may produce new *combinations* of genes, but not new *types* of genes

n. **False,** because directional selection tends to eliminate alleles with deleterious effects on the phenotype, counteracting the effect of the mutation that produces them

o. **False,** because initial allele frequencies are irrelevant; equilibrium allele frequencies depend solely upon the forward and reverse mutation rates:

$$\hat{p} = v/(u + v), \text{ and } \hat{q} = u/(u + v)$$

p. **False,** because with unidirectional recurrent mutation, one allele (the one that mutates) will eventually be eliminated from the population

19. a. $\hat{q} = \sqrt{u/s} = \sqrt{0.9} \times 10^{-5}/1 = \sqrt{0.000009/1} = 0.003$

    q^2 = frequency of hh = 0.000009. In a population of 10 million, 90 individuals (10,000,000 × 0.000009) will have Hurler syndrome.

   b. $\hat{q} = \sqrt{0.9 \times 10^{-5}/0.5} = \sqrt{0.000009/0.5} = \sqrt{0.000018} = 0.0042426$

    q^2 = frequency of hh = 0.0000179. In a population of 10 million, 179 individuals will have Hurler syndrome.

   c. $t = \dfrac{1}{q_{t2}} - \dfrac{1}{q_{t1}}$

   Because $\hat{q}$ before the medical advance = 0.003 = about 1/333, and $\hat{q}$ after the advance = 0.0042426 = about 1/236, t = 333 − 236 = 97 generations.

13
Speciation

Problems

1. In 1951, Theodosious Dobzhansky defined *species* as follows: groups of populations, the gene exchange between which is limited or prevented in nature by one or a combination of several reproductive mechanisms. Does this definition best describe morphological species (structural differences), biological species (reproductive isolation), or evolutionary species (examination of the fossil record)?

2. Three allopatric populations of plants exist in nature, each differing in average plant height, flower color, and leaf shape. In the greenhouse, a botanist discovers that matings between plants from populations 1 and 2 and between 2 and 3 produce healthy but sterile offspring, whereas matings between plants from populations 1 and 3 produce healthy, fertile offspring. How many morphospecies (species distinguished by phenotypic differences) and how many biospecies (species distinguished by reproductive behavior) are represented? Explain.

3. *Drosophila pachea* exists only in the Sonora Desert and breeds exclusively in the stems of the senita cactus (*Lophocereus schottii*). Two other *Drosophila* species breed in the fruits of this cactus, but not in the stems. The stems produce a sterol compound that is lethal to the cactus-fruit-breeding species but is necessary for breeding to occur in *D. pachea* because breeding will not occur in the lab unless a piece of stem or the sterol is added to the culture medium. What type of isolating mechanism is acting here between *D. pachea* and the other species?

4. Autotriploids have very low fertility because the odd number of chromosomes decreases the probability of forming balanced ($1n$) gametes. The chromosome numbers of three autotriploid plants are 9 for plant 1, 12 for plant 2, and 15 for plant 3.
   a. For each, determine the probability of producing a balanced gamete.
   b. If plant 1 attempts self-fertilization, what is the probability of a fertilization producing a chromosomally balanced (i.e., viable) offspring?

5. Two species of plant exist in the same general region. Each species has 9 pairs of chromosomes ($2n = 18$). Occasionally, in regions where their ranges overlap, hybrids between these species occur. However, the hybrids are sterile because the chromosomes do not pair properly during meiosis. Explain how rapid speciation could occur in this situation.

6. American cultivated cotton (*Gossypium hirsutum*) has 52 chromosomes and arose as an allotetraploid produced from a cross of Asiatic cotton (*G. arboreum*) and another American variety, *G. thurberi*. Both *G. arboreum* and *G. thurberi* have chromosome numbers of 26, so *G. hirsutum* must have been produced by chromosome doubling of the F_1.
   a. What is the basic ($1n$) number for cotton?
   b. What would be the chromosome number and level of ploidy of the offspring of a cross between American cultivated and Asiatic cotton?
   c. Would the offspring of this cross be fertile?

7. Three closely related plant species (species R, S, and T) each have a diploid number of 8 chromosomes. Diagram the steps involved in the rapid evolution of a fertile allohexaploid species derived from species R, S, and T. Begin with species R mating with species S.

8. Two species that exist within the mantid genus *Liturgousa* are *L. maya* ($2n = 18$ in females) and *L. cursor* ($2n = 34$ in females). Males in each species have one less chromosome per cell because they are XO. Comparison of the chromosomes in males of the species reveals the following:

SPECIES	MALE DIPLOID NUMBER OF CHROMOSOMES	TOTAL NUMBER OF CHROMOSOME ARMS	RELATIVE AMOUNT OF DNA PER NUCLEUS
L. maya	17 (XO)	33	1.00
L. cursor	33 (XO)	33	0.94

   a. Is it likely that *L. cursor* arose by polyploidy from *L. maya*? If not, what has happened?

   b. Which species is more likely to have acrocentric chromosomes and which is more likely to have metacentric chromosomes?

9. Several chromosomal differences exist between various species of mantids in the genus *Ameles*. Comparisons of chromosome and DNA content among three species resulted in the following data:

SPECIES	MALE DIPLOID NUMBER OF CHROMOSOMES	CHROMOSOMES METACENTRIC	ACROCENTRIC	RELATIVE MEAN DNA/ SPERMATID
Ameles ameles	19	11	8	2.07
A. heldreichi	27	3	24	1.84
A. andrei	28	2	26	1.71

Assume *A. andrei* is the oldest species. Give the mechanisms responsible for the chromosome changes, and explain the evolutionary relationship of the three species.

10. Mitochondrial DNA (mtDNA) from seven geographically distinct populations of a species is analyzed by restriction endonuclease digestion. Gel electrophoresis shows the following patterns of mtDNA fragments from the seven populations:

POPULATION	mtDNA FRAGMENTS (KILOBASES)
A	4.7, 13.0
B	2.5, 15.2
C	4.7, 6.1, 6.9
D	7.6, 10.1
E	2.2, 2.5, 13.0
F	6.1, 11.6
G	2.5, 5.1, 10.1

Population B is known to be the oldest population. What is the most likely evolutionary relationship of these seven populations?

Challenge Problems

11. The cotton species *Gossypium barbadense* has 13 large pairs and 13 small pairs of chromosomes in each diploid cell. *G. thurberi* has 13 small pairs of chromosomes per diploid cell, and *G. herbaceum* has 13 large pairs of chromosomes per diploid cell. During meiosis in various hybrids between these species, the following chromosome pairings occur:

MATING	CHROMOSOME PAIRING AT METAPHASE I IN THE HYBRIDS
G. barbadense × *G. thurberi*	13 small bivalents and 13 large univalents
G. barbadense × *G. herbaceum*	13 large bivalents and 13 small univalents
G. thurberi × *G. herbaceum*	13 large univalents and 13 small univalents

What are the evolutionary relationships among these three cotton species?

12. In humans, the following minimum number of DNA nucleotide substitutions accounts for the amino acid differences between the alpha-, beta-, and gamma-hemoglobin polypeptides:

    beta-gamma = 47 DNA nucleotide differences

    beta-alpha = 111 DNA nucleotide differences

    alpha-gamma = 114 DNA nucleotide differences

The gene for alpha-hemoglobin is in chromosome 16, and the beta and gamma genes are tightly linked in chromosome 11. Explain the evolutionary history of these three genes in humans.

Team Problems

13. Two wild species of tobacco, *Nicotiana sylvestris* and *N. tomentosiformis*, are 2n = 24. Cultivated tobacco, *N. tabacum*, is 2n = 48. *N. tabacum* haploid plants are sterile, with 24 univalents occurring during metaphase I of meiosis. Matings among the three diploid species reveal the following patterns of chromosome pairings during meiosis in the hybrids:

MATING	CHROMOSOME PAIRING AT METAPHASE I IN THE HYBRIDS
N. sylvestris × *N. tomentosiformis*	24 univalents
N. sylvestris × *N. tabacum*	12 bivalents and 12 univalents
N. tomentosiformis × *N. tabacum*	12 bivalents and 12 univalents

What is the probable evolutionary relationship among the three species of tobacco?

14. The number of DNA nucleotide base differences in the beta-globin gene region between humans, chimpanzees, gorillas, and orangutans has been estimated as follows:

    human-chimpanzee = 115 nucleotide differences

    human-gorilla = 125 nucleotide differences

    human-orangutan = 237 nucleotide differences

    chimpanzee-gorilla = 154 nucleotide differences

    chimpanzee-orangutan = 266 nucleotide differences

    gorilla-orangutan = 260 nucleotide differences

    Determine the most likely evolutionary relationship among these species using the beta-globin gene data above.

15. In northwestern Mexico, six chromsomal "races" of the pocket gopher (*Perognathus goldmani*) exist in adjacent populations. Diagrams of the chromosomes that differ among these races (designated by Greek letters) and the theoretical ancestral chromosomal arrangement are given below. Propose a sequence of changes in these chromosomes that explains the evolutionary relationships of these races.

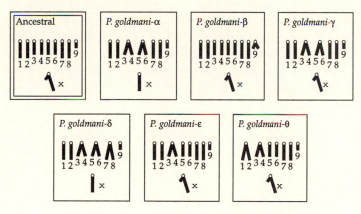

Solutions

1. This definition best describes biological species because it stresses reproductive isolation and does not require morphological differentiation or examination of the fossil record in determining species relationships.

2. Plants in the three populations can be considered three morphologically distinct species. However, only two biospecies are present. Because individuals from populations 1 and 3 produce fertile offspring when intercrossed, these populations constitute the same biospecies (species "A") even though they are morphologically distinct. The two populations of species A may be regarded as different "races" or "subspecies," just as black and white humans are different races but interbreed to produce fertile offspring. Individuals from population 2, however, produce sterile hybrids when mated with population 1 or 3 individuals. Therefore, population 2 is a different species (species "B") than species A because reproductive isolation exists, just as humans and chimpanzees are different species.

3. This is an example of habitat (ecological) isolation between *D. pachea* and the other two species. The sterol compound in the stem is necessary for *D. pachea* to breed but is lethal to the other species.

4. For triploids, the probability that a gamete will receive a balanced haploid set of one of each chromosome type is $(1/2)^x$, where x = the haploid number of chromosomes.

   a. For plant 1 ($3n = 9$), three types of chromosomes are present ($n = 3$). The probability that any particular gamete is chromosomally balanced (having one of each of the three types of chromosomes) is $(1/2)^3 = 1/8$ or 12.5%. For plant 2 ($3n = 12$; $n = 4$), and $(1/2)^4 = 1/16$ or 6.25%. For plant 3 ($3n = 15$; $n = 5$), the chance of a gamete being chromosomally balanced is $(1/2)^5 = 1/32$ or 3.125%.

   b. Using the product rule (because the chance of two chromosomally balanced gametes finding each other through random fertilization is a matter of chance), $1/8 \times 1/8 = 1/64$ or 1.5625%.

5. Let's symbolize the chromosome complements for species *A* as $2n = 18 = AA$, for species *B* as $2n = 18 = BB$, and for the sterile hybrids as $2n = 18 = AB$ (each *A* and *B* symbol represents 9 chromosomes). If somatic doubling of the chromosomes in a hybrid occurred, it would become a fertile allotetraploid: $4n = 36 = AABB$. Self-fertilization would quickly allow a population of this new "species" to become established if favored by natural selection. The allotetraploid species would be reproductively isolated from both parental species because $2n\ AA \times 4n\ AABB \rightarrow 3n\ AAB$ (sterile) and $2n\ BB \times 4n\ AABB \rightarrow 3n\ ABB$ (sterile). Premating isolating mechanisms would then evolve to reduce interspecific pollination.

6. a. $1n = 13$. Both *G. arboreum* and *G. thurberi* must be diploid.

   b. 39 chromosomes, $3n$, because a gamete from American cultivated would contain 26 chromosomes and a gamete from Asiatic has 13 chromosomes

   c. no, because it is triploid

7.
```
           species R              species S
             RR          ×          SS
         n = 4 (fertile)       n = 4 (fertile)
          x = 2 sets      ↓     x = 2 sets
           diploid               diploid

                    RS
               n = 4 (sterile)
                 x = 2 sets
                allodiploid

   CHROMOSOMAL
    DOUBLING    ↓
             species RS             species T
               RRSS        ×          TT
           n = 8 (fertile)       n = 4 (fertile)
             x = 4 sets     ↓      x = 2 sets
          allotetraploid           diploid

                    RST
               n = 6 (sterile)
                 x = 3 sets
               allotriploid
```

CHROMOSOMAL
DOUBLING $\downarrow$

species *RST*

RRSTT

$n = 12$ (fertile)

$x = 6$ sets

allohaxaploid

8. a. The two species are not related by polyploidy because they have the same amount of DNA, even though the number of chromosomes is twice as great in *L. cursor*. In all probability, Robertsonian (centric fusion) translocations occurred to convert the *L. cursor* type chromosomes into the *L. maya* type chromosomes.

    b. The chromosomes of *L. maya* are metacentric, and those of *L. cursor* are acrocentric.

9. The chromosomes evolved in *A. heldreichi* and *A. ameles* by centric fusion (Robertsonian) translocation of some of the metacentrics in the original species, *A. andrei*. If a straight-line sequence of speciation occurred in these species, the relationship is *A. andrei* → *A. heldreichi* → *A. ameles*. Fusion of two acrocentrics produced one of the metacentrics in *A. heldreichi*, and fusion of 18 acrocentrics produced 9 of the metacentrics in *A. ameles*.

10. Analyze the fragments as follows:

    because 15.2 = 10.1 + 5.1, B → G

    because 7.6 = 5.1 + 2.5, G → D

    because 15.2 = 13.0 + 2.2, B → E

    because 4.7 = 2.2 + 2.5, E → A

    because 13.0 = 6.1 + 6.9, A → C

    because 11.6 = 6.9 + 4.7, C → F

    Thus, the most likely evolutionary relationship of the seven populations is:

    D ← G ← B → E → A → C → F

11. In all probability, *G. barbadense* is an allopolyploid species that arose from the mating between *G. thurberi* and *G. herbaceum*. From the matings, it appears that the 13 large chromosomes in *G. barbadense* and *G. herbaceum* are similar because they synapse during meiosis. Likewise, the 13 small chromosomes in *G. barbadense* and *G. thurberi* are similar in origin.

12. Assume that fewer nucleotide differences means a closer ancestral relationship (i.e., the common ancestor existed more recently) and vice versa and that mutations occur at a roughly equal rate over time. Using the numbers of DNA nucleotide differences between the hemoglobin types, the following phylogenetic relationship may be inferred:

    Dividing the phylogeny into regions x, y, and z, we can determine the number of nucleotide substitutions that occurred in each of its branches by using some simple algebra.

    I: $x + y = 47$

    II: $x + z = 111$

    III: $y + z = 114$

If we subtract II from III, we get IV: $y - x = 3$. Adding IV and I, we get $2y = 50$, $y = 25$. If $y = 25$, from I: $x + 25 = 47$, $x = 22$, and from III: $25 + z = 114$, $z = 89$.

So the phylogeny makes sense because:

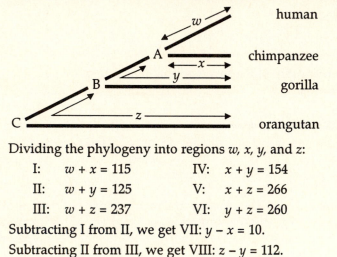

Because the alpha gene is on chromosome 16 and the beta and gamma genes are closely linked on chromosome 11, the putative ancestral gene B was present in chromosome 16. It duplicated, and the duplicate was translocated onto chromosome 11. Gene B evolved into the alpha gene. The duplicate on chromosome 11 evolved into gene A, then duplicated, but both copies remained tightly linked. One copy evolved into the beta gene, and the other became the gamma gene.

13. *N. tomentosiformis* and *N. sylvestris* are probably distantly related because none of their chromosomes synapse during meiosis. *N. tabacum* is probably an allotetraploid derived from the mating of *N. tomentosiformis* × *N. sylvestris* because half the *N. tabacum* chromosomes synapse with the 12 *N. tomentosiformis* chromosomes and half the *N. tabacum* chromosomes synapse with the 12 *N. sylvestris* chromosomes. Also, a haploid *N. tabacum* cell has chromosomes that behave in a fashion similar to those in the hybrid between *N. sylvestris* and *N. tomentosiformis*.

14. Using the numbers of DNA nucleotide differences between human-chimp, human-gorilla, and human-orangutan, the following phylogenetic relationship is inferred:

Dividing the phylogeny into regions w, x, y, and z:

I: $\quad w + x = 115$	IV: $\quad x + y = 154$
II: $\quad w + y = 125$	V: $\quad x + z = 266$
III: $\quad w + z = 237$	VI: $\quad y + z = 260$

Subtracting I from II, we get VII: $y - x = 10$.

Subtracting II from III, we get VIII: $z - y = 112$.

Adding IV and VII, we get $2y = 164$, $y = 82$.

Adding VI and VIII, we get $2z = 372$, $z = 186$.

From II, $w + y = 125$, $w = 125 - 82 = 43$.

From I, $w + x = 115$, $x = 115 - 43 = 72$.

So the phylogeny makes sense because:

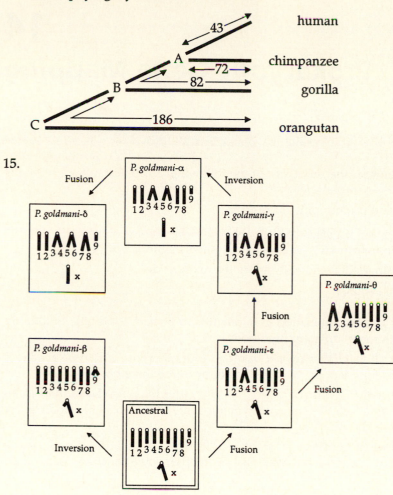

15.

14
Prokaryotic Gene Mapping

Problems

1. What is the best genetic notation for the genotypes of the following *E. coli* mutants with the following phenotypes?

   a. auxotrophic for arginine and prototrophic for histidine

   b. able to grow on xylose alone and unable to grow on lactose alone

   c. requires the vitamin biotin for growth and doesn't require exogenous purines

   d. sensitive to the antibiotic ampicillin and resistant to the toxin azide

   e. can use glucose as a carbon source

2. When studying gene transfer and recombination in *E. coli*, it is important to be able to identify, preserve, and grow *E. coli* with many different mutations.

   a. How would you select for all *E. coli* that are *mal⁻ tonˢ trp⁻* (but otherwise wild type at all loci) among *E. coli* that are wild type for one, two, and/or three of the three genes?

   b. How would you select for all *E. coli* that are *mal⁻ tonˢ trp⁻* (but may also be mutant at the *his*, *arg*, and *cys* loci) among *E. coli* that are wild type for some of the six genes and mutant for others?

   c. How would you determine the genotype (at three loci) for all *E. coli* that have one, two, or three of the following: *mal⁻ tonˢ trp⁻*, but may also have other mutations that make them auxotrophic?

3. One way to isolate auxotrophic mutations for amino acid biosynthesis in *Neurospora crassa* would be to mutagenize conidia, cross with wild-type *Neurospora* of the opposite mating type, dissect out ascospores, and transfer one by one to culture tubes containing:

   a. one of the intermediates in a biosynthetic pathway. If the conidia give rise to growth, they are not deficient in any enzyme working downstream (after) the intermediate in the biosynthetic pathway. If the conidia do not grow, they are auxotrophs for some enzyme earlier in the pathway.

   b. minimal medium plus one of the 20 essential amino acids. If growth is seen in all of these 20 tubes, transfer to complete medium. If no growth is seen in complete medium, transfer to minimal medium, then transfer all conidia that grow in complete medium to tubes marked with the amino acids they grew in. These are the auxotrophs for the amino acids.

   c. first complete medium, then transferring conidia from each tube of complete medium that shows growth to a tube containing minimal medium, then, for each of the tubes that didn't show growth in the minimal medium, transferring conidia from the first tubes to 20 tubes each containing minimal medium plus one of the 20 essential amino acids.

d. first minimal medium, then transferring conidia from each tube of minimal medium that shows growth to a tube containing complete medium, then, for each of the tubes that shows growth in the complete medium, transferring conidia from the minimal medium tubes to 20 tubes each containing minimal medium plus one of the 20 essential amino acids.

e. none of the above.

4. *Bacillus subtilis* undergoes natural transformation at an unusually high frequency, about 1 transformation per 1,000 cells per gene.

a. If you plate 10,000,000 competent *Bacillus subtilis* that are *lac⁻ arg⁻* that have been mixed with excess wild-type *Bacillus subtilis* DNA, how many colonies would you expect to find on a plate without arginine and lactose as the carbon source?

b. What would you conclude if you found almost 10^4 colonies?

c. Draw the recombination intermediate that gave rise to the *lac⁺ arg⁺* cells.

5. Bacterial conjugation and natural transformation share some common features.

a. Distinguish between the mechanism of *gene transfer* in conjugation and natural transformation.

b. Distinguish between the mechanism of *genetic recombination* in conjugation and transformation.

6. Conjugations are set up between *E. coli* of the genotypes indicated below. Bacteria are incubated for 10 minutes and then plated on minimal medium with a concentration of azide sufficient to kill azide-sensitive *E. coli*. Predict how many *Hfr*, *F⁺*, and *F⁻* colonies (many, few, none) will appear on the azide plates *and* how they could come about. Note: the *purE* locus is very near the origin of the integrated F factor, *HfrC*. Also, the mutations in *purE* and *azi* are both large deletion mutations within the genes.

a. *HfrC aziˢ purE⁺ × F⁻ aziʳ purE⁻*

b. *HfrC aziˢ purE⁻ × F⁻ aziʳ purE⁺*

c. *HfrC aziˢ purE⁺ × F⁺ aziʳ purE⁻*

d. *F⁺ aziˢ purE⁺ × F⁻ aziʳ purE⁻*

7. You have been given three *Hfr* strains of *E. coli* that show the following marker order in an interrupted mating experiment:

Hfr strain 1: BEFD

Hfr strain 2: EBAC

Hfr strain 3: DCAB

Draw a schematic diagram of the *E. coli* chromosome with respect to these markers, placing on the diagram the locations of integration of the sex factor (F) in the genome in the three strains (label "F_1," "F_2," etc.). Indicate directionality of the sex factor in the standard manner.

8. The graph below shows the results of an interrupted mating experiment for a close relative of *E. coli*. Using the data, propose a map of these genes. Indicate distances between all loci with the appropriate units.

   HfrG11 azi^s × *F$^-$ mal arg$^-$ xyl$^-$ pyr$^-$ met$^-$ pur$^-$*

   Plating is on complete medium except for the following:
   No arginine is provided.
   Maltose is the primary carbon source.
   Azide is at selective concentration.

   Replica plates were then made to assess each colony's genotype at the *xyl$^-$ pyr$^-$ met$^-$* and *pur$^-$* loci.

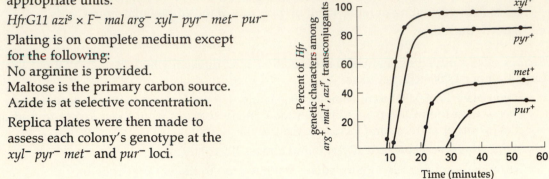

9. In a general transduction experiment, P1 phage are grown on *trp$^+$ lac$^+$ amp^R tet^S E. coli* and are used to transduce *E. coli* that are *trp$^-$ lac$^-$ amp^S tet^R*. What would you supplement minimal media (agar, salts) with to identify the *trp$^+$ lac$^+$ amp^R tet^R* transductants that you plate on these media?

10. In a transduction experiment, closely spaced genes can be mapped. Using the general transduction experiment data provided in the table below, generate an ordered map of the three *E. coli* genes, *a*, *b*, and *c*, and indicate the approximate spacing of the genes.

SELECTED MARKER	NUMBER	UNSELECTED MARKER	NUMBER
a^+	200	b^+	60
		b^-	140
		c^+	20
		c^-	180
c^+	250	a^+	25
		a^-	225
		b^+	5
		b^-	245

11. Transduction is one of the main ways that DNA transfer occurs in bacteria.

    a. Define the essential difference between general and specialized transduction.

    b. How are specialized transduction and sexduction similar?

12. Describe in detail how λ*d bio$^+$* could be generated and how it works to produce a high-frequency transducing (HFT) lysate when used transducing *bio$^-$ E. coli*.

13. What assumption does one make when using recombination of bacteriophage genes as a measure of genetic distance between phage genes?

14. Draw a schematic diagram showing how recombination occurs in the bacteriophages such as T4.

15. What is the structure shown below?

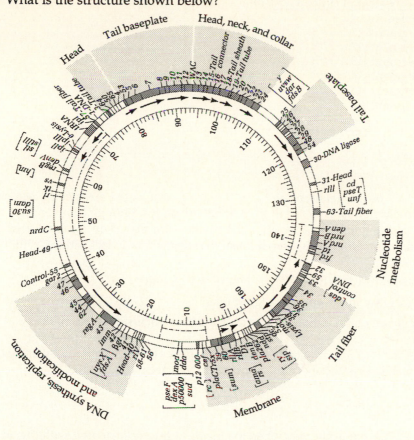

16. Deletion mapping is an invaluable tool for speeding the process of mutation mapping. The following crosses were done to obtain an initial mapping of four new phage T4 *rII* mutations. From the data provided, place the mutants on the map of the *rII* A and B loci shown below. A "+" indicates that the two strains when used to coinfect bacteria can recombine to wild type. A "−" means they cannot.

				TEST STRAIN				
rII MUTANT	*r+*	*r1272*	*r1241*	*rJ3*	*rPT1*	*rPB242*	*rA105*	*r638*
1001	+	−	−	−	+	+	+	+
1002	+	−	−	−	−	−	+	+
1003	+	−	+	+	+	+	+	+
1004	+	−	−	−	−	−	−	+

Phage
T4   —|————|————|————|————|————|————|————|—
rIIA,B
map   A1   A2   A3   A4   A5   A6   B

TEST STRAIN	REGION DELETED (INCLUSIVE)
r^+	no deletion
r1272	rII segment A1 to rIIB
r1241	rII segment A2 to rIIB
rJ3	rII segment A3 to rIIB
rPT1	rII segment A4 to rIIB
rPB241	rII segment A5 to rIIB
rA105	rII segment A6 to rIIB
r638	rIIB

Challenge Problems

17. A lac^+ Hfr strain of *E. coli* of the genotype str^S i^+ o^+ z^+ y^+ (grown in the absence of lactose) is mated to a lac^- F^- recipient that is str^R i^- o^+ z^- y^-. After allowing just enough time for the transfer of the genes indicated above (except for str^S), the bacteria are plated on minimal medium, arabinose and streptomycin, but, again, no lactose. Transient beta-galactosidase activity is observed in some colonies.

    a. Explain how beta-galactosidase can be made by the str^R colonies.

    b. Why is beta-galactosidase made only for a short time?

    c. How would you explain colonies that had persistent beta-galactosidase activity?

18. Propose a way to generate a *stable* merozygote *E. coli* for the genes of the *his* operon by using conjugation experiments and selection only, not recombinant DNA technology.

Team Problems

19. You have selected for *E. coli* that grow only when the medium is supplemented with the amino acid cysteine and believe you have found a new gene necessary for cysteine biosynthesis. Because you know that other "cys" genes have been found, you decide to map your newly discovered gene, *cysX*.

    You have the following two strains of bacteria:

    Hfr $cysX^-$ gal^+ his^+ lac^+ pur^+ trp^+ ton^s

    F^- $cysX^+$ gal^- his^- lac^- pur^- trp^- ton^R

    Key to mutations:

    $cysX^-$ indicates the inability to make cysteine.

    lac^- and gal^- indicate the inability to use the sugar lactose or galactose, respectively, as the sole energy source.

    ton^s indicates sensitivity to the phage T1.

    his^-, pur^-, and trp^- indicate auxotrophy for histidine, purines, and tryptophan, respectively.

    a. What should you supplement your minimal media with to recover *all* $cysX^+$, lac^+, and ton^R recombinants resulting from a cross of the *Hfr* and F^- strains? What should *not* be present? Be sure to indicate whether or not you would include glucose.

    b. What classes of recombinants would be selected on plates containing only minimal media, glucose, histidine, purines, and T1 phage? Include only genotypes that will be unequivocally identified.

       You perform an interrupted mating experiment with the *Hfr* and F^- strains. Ton^R colonies with each of the following phenotypes first appear at the times (in minutes) shown below:

PHENOTYPE	MINUTES
cys^+	0
cys^-	25
gal^+	17
his^+	38.5
lac^+	9.5
pur^+	14.5
trp^+	25

c. Sketch a bacterial chromosome and indicate the relative position of each of the markers and the *cysX* gene *in the Hfr strain*. Label distances between adjacent markers.

Based on the results above, you decide to do the higher resolution mapping of the *cysX* gene using P1 cotransduction. P1 phage grown on an *E. coli* strain of the genotype $cysX^- ton^R trp^+$ are used to infect a recipient strain of *E. coli* that is $cysX^+ ton^S trp^-$. These bacteria are then plated on complete media in the presence of T1 phage. Two hundred ton^R transductants are scored for their genotype with respect to the other two loci. The results are as follows:

GENOTYPE	NUMBER OF COLONIES
$cysX^+ trp^+$	10
$cysX^+ trp^-$	10
$cysX^- trp^+$	180
$cysX^- trp^-$	0

d. What is the recombination frequency between *trp* and *cysX*?

e. What is the recombination frequency between *ton* and *cysX*?

f. What is the order of the three genes? Indicate the distance between each gene.

20. Based on the following P1 transduction data, what is the order of *aroA*, *cmlB*, and *pyrD* on the *E. coli* chromosome?

DONOR	RECIPIENT	SELECTED MARKER	UNSELECTED MARKER*	FREQUENCY
aroA pyrD+	*aroA+ pyrD*	*pyrD+*	*aroA*	10/200
aroA+ cmlB	*aroA cmlB+*	*aroA+*	*cmlB*	52/200
cmlB pyrD+	*cmlB+ pyrD*	*pyrD+*	*cmlB*	27/50

*Ten of the 200 *pyrD+* colonies in the first transduction were also *aroA*.

Solutions

1. a. $arg^- his^+$
   b. $xyl^+ lac^-$
   c. $bio^- pur^+$
   d. $amp^S azi^r$
   e. wild type for most or all of the genes encoding enzymes of the glycolysis pathway

2. a. Plate the bacteria on minimal medium with glucose as the primary carbon source, supplemented by amino acid tryptophan, and free of phage T1. Then replica plate the colonies that grow up on these plates onto those that have maltose as the sole carbon source, no supplemental tryptophan, and abundant T1 bacteriophage. Those that *don't grow* on the selective medium are the ones that you want, mal^- ton^s trp^-. Recover these by returning to the colonies on the complete medium that correspond to the colonies that didn't grow on the selective plate.

    b. Plate bacteria on medium containing complete medium (which includes glucose, basic salts, all essential amino acids, and the vitamins *E. coli* need for optimal growth) and no phage T1. This will select for *E. coli* that have the genes necessary for viability in the best growth conditions. Included in this are those that are his^-, arg^-, cys^-, mal^-, ton^s and/or trp^-. Next replica plate these bacteria to plates with complete medium that lacks one of these amino acids, has only maltose as the carbon source, or contains bacteriophage. These plates will type the bacteria at the specific loci that are expected to be mutated in some of the colonies.

    c. Do as in the first part of part a, but instead of plating the *E. coli* onto plates that select for all three traits, plate onto plates that select for only one of the factors at a time. Repeat the replica plating until all three genotypes have been separately tested.

3. The correct answer is c. This procedure allows one first to gather prototrophs *plus a broad array of auxotrophs* (by growing organisms on complete medium), then to identify the auxotrophs (by growing organisms on minimal medium), and then to test each auxotroph for its particular deficiency (by growing organisms on supplemented minimal medium).

4. a. You would normally expect $1/1{,}000 \times 1/1{,}000$ probability for the transformation of two genes in *Bacillus subtilis* by natural transformation. This joint probability is the product of the two independent probabilities. The number expected is this probability times the total number of bacteria screened, 10,000,000 in this example. $10^{-6} \times 10^7 = 10$. You would expect 10 colonies on the lactose/no arginine plates.

    b. If you found 10,000 colonies (or anything approaching that), you could conclude the two genes are closely linked and therefore often *cotransform* a cell.

    c.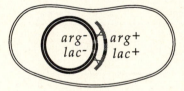

5. a. In both conjugation and transformation, gene transfer involves the entry of single-stranded "donor" DNA into a recipient cell. In conjugation, however, the transfer is direct, from another living cell through a pilus (a conduit for cell-to-cell DNA transfer). In conjugation, the cell donating the DNA replicates the strand left behind so that the double-stranded *F* factor is regenerated. In conjugation, the transferred DNA always begins with the same nicked portion of the *F* factor. If the *F* factor is completely transferred during conjugation, the recipient cell becomes able to transfer DNA to other recipient cells. In natural transformation, DNA is not threaded through a pilus, but enters the cells through pores in a process that is still not well understood.

    b. Recombination following natural gene uptake in *Bacillus subtilis* involves the formation of triplex DNA by strand displacement and double recombination events. Subsequent replication converts the heteroduplex DNA to homoduplex DNA. Following *E. coli* conjugation, the donor DNA that has successfully transferred to the host cell is replicated and often recircularized. The single-stranded or replicated dsDNA can find homologous sequences in the recipient's genome and recombine there, changing the genotype of the recipient cell.

6. Each of the following answers describes various types of bacterial colonies, including those that are *not* produced by simple conjugation, but rather by a combination of mutation, *F⁻* plasmid mobility, and natural transformation.

    a. *Many F⁻* colonies by conjugation-mediated gene transfer and recombination-mediated conversion of *purE⁻* to *purE⁺*.

        Few if any *Hfr* colonies (occurring at only *E. coli*'s natural transformation frequency).

        Very few if any *F⁺* colonies (these require that the *F* plasmid excise itself from *E. coli* genome imprecisely [bringing the *purE* gene with it] or for the coincidence of conjugation [conversion of *F⁻* to *F⁺*] and transformation of the *F⁻* strain with *purE⁻*).

    b. Lawn of *F⁻* colonies because they already are purine prototrophs and azide resistant. Very few *Hfr*. They would survive only if cotransformed for genes. No *F⁺* colonies because that would require both being cotransformed for the two genes and having the F plasmid be excised from the *E. coli* genome.

    c. Few *F⁻* because they will occur only by natural transformation with *purE⁺*.

        Few *Hfr* because they require natural transformation either by itself or with the concurrent integration of an episomal *F⁻* plasmid. Few *F⁺* because they require natural transformation with *purE⁺*.

    d. Few *F⁻* because all pathways require a gene to be repaired by natural transformation.

        No *Hfr* because both transformation and integration of the F factor would be required.

        Few *F⁺* because they would require natural transformation by itself or with standard F-plasmid transfer.

7. The map below (or its mirror image) can be produced by aligning the marker orders obtained in the interrupted mating experiment to give a contiguous—and in this case, circular—sequence of markers.

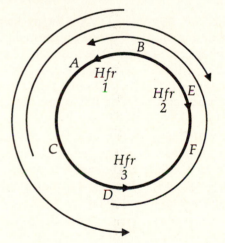

8. In conjugation mapping, distances between markers are determined by the differences in time of appearance of those markers. The time of appearance for a marker can be determined from the graph given. Where the line representing the percentage of *mal⁺ arg⁺* and *aziʳ* emerges from the *x* axis for the marker in question is the time of appearance for that marker. For example, colonies with the ability to live on xylose as a carbon source are first seen at 12 minutes after the beginning of the mating. Ability to synthesize pyrimidines is seen shortly thereafter, at 14 minutes. Hence, the loci *xyl* and *pyr* are 2 minutes apart.

How high the percentage of a given marker type rises is also a function of the distance from that marker to the originally selected markers (*mal⁺* and *arg*) in the vicinity being mapped. The *xyl* line, for example, reaches a plateau at 90% suggesting it is almost always recombined with the *mal* locus as long as it has successfully been transferred into the cell (which is what *hasn't* happened at 10 minutes). The *pur* gene never rises higher than 35%, reflecting that there are many times when a crossover even separates the two loci because of their distance apart. Remember that all the bacteria being tested in this experiment are known to be *mal⁻*. *pur⁺ mal⁻* recombinants would not be detected in this experiment.

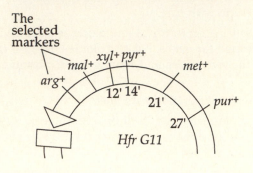

9. lactose, tetracycline, ampicillin; no glucose and no tryptophan

10. cotransduction frequency (*a–b*) = 60/(60 + 140) = 30%
cotransduction frequency (*a–c*) = 20/(20 + 180) = 10%
cotransduction frequency (*c–a*) = 25/(25 + 225) = 10%
cotransduction frequency (*c–b*) = 5/(5 + 245) = 2%

The map (units in cotransduction frequencies):   *b* – 30% —— *a* —— 10% —— *c*

11. a. During general transduction, a random piece of the bacterial genomic DNA is mispackaged into a phage head. This nonphage DNA can be donated to another bacteria because all the necessary cell-recognition molecules are part of the phage coat and tail. The fragment of bacterial DNA from the previous host can then recombine with the new host genome and, in some instances, replace nonfunctional genes in the new host. The hallmark of general transduction is that each transducing phage can rescue a broad array of mutant bacteria, and the host cell is not harmed because there is no lambda phage DNA in the phage heads.

   Specialized transduction occurs when a phage excises itself from its host genome *imprecisely*, resulting in nearby DNA being substituted for some other bacteriophage genes. Because the genes that can be accidentally excised are only a subset of bacterial genome—they must be genes flanking the integration site—a hallmark of specialized transduction is that only a limited range of mutations can be rescued by the specialized transducing phage.

   b. Both specialized transduction and sexduction occur with imprecise excision from the bacterial genome, are capable of mediating a high frequency of recombination, but involve the transfer of a limited array of genes to the new hosts.

12. Lambda phage integrates at the *att* site in the *E. coli* genome. When it loops out, sometimes the excision is not precise. That is, flanking *E. coli* DNA may be removed, and some important lambda genes may be left behind. This imprecisely excised lambda can, however, be packaged because all the lambda genes are present either in the bacterial genome or in the partially deleted lambda genome.

   Most of the phage produced from the initial infection are wild type with very few deleted/transducing forms. Therefore, the lysate is called low-frequency transducing.

When these deleted lambda are used to infect cells in a second infection, one of two types of transductants will be produced. (1) Wild-type lambda phage in the lysate will integrate at its normal *att* λ site, and then the λd *bio*+ phage integrates within the lambda prophage, rather than within another bacterial gene. As a consequence of this, the lambda genes deleted in the λd *bio*+ phage are available from the wild-type lambda genome. The *bio*+ gene is also incorporated into the host bacterial genome, creating a merozygous state for the *bio*+ gene. The transductant is not stable, however, because all the lambda genes required for lysis are present and can be used; these transductants give rise to the high-frequency transducing lysate.

(2) Alternatively, only λd *bio*+ phage may infect a bacterium. The *bio*+ gene is able to mediate integration of the phage at the defective *bio* gene in the bacterium. If this occurs, the bacterium is both stably transduced to *bio*+ and free from the instability inherent in lysogeny. Progeny of this kind of tranductant will also have a stable *bio*+ genotype.

13. That crossing-over happens randomly over the entirety of a chromosome and that recombination frequency is a good measure of map distance between genes, in this case, phage genes.

14. a. Coinfect bacteria with the two parental phages, *h*+ *r* and *h r*+

   b. Replication of phage chromosomes in cell

   c. Recombination between some parental chromosomes

   d. Phage assembly, bacterial lysis, and release of progeny phages

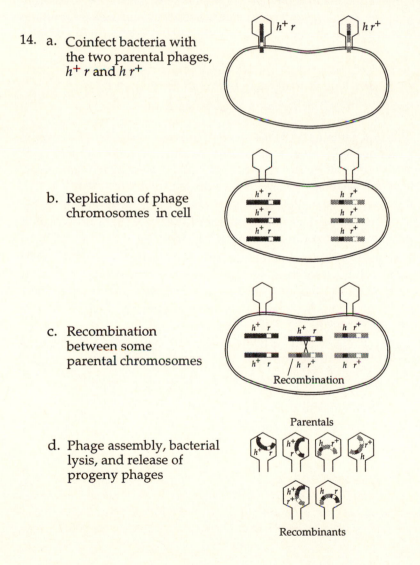

15. The genome of a bacteriophage, specifically the T4 genome. Note that the genome is very small (166 kb) and that genes present are largely encoding the coat proteins. Few genes encoding RNAs or enzymes necessary for replication, transcription, and translation and no genes for basic energy metabolism are present. Unlike in a mitochondrial or chloroplast genome, no genes encoding electron transport proteins or photosynthetic enzymes are present. Note the location of *rIIA* and *rIIB* at location marked "0."

16.

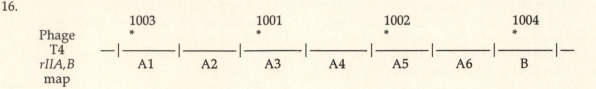

17. a. A good copy of *lacZ* is provided by the *Hfr* strain. The *F⁻* recipient does not have preexisting repressor protein. Though a functional copy of the repressor gene is also provided, transcription (and translation) of the *lacZ* gene will occur before sufficient repressor protein has been produced to stop transcription at the *lac* operon. mRNA and beta-gal protein are sufficiently stable to see activity.

    b. The *Hfr* strain provides a good *i* (repressor) gene. As repressor protein accumulates, *lac* operon transcription will be reduced as long as there is no lactose in the medium.

    c. The repressor gene may not have been transferred or was lost early on.

18. One way is to generate an *F'* plasmid containing the *his* operon (and perhaps some of the genes in the vicinity). Consulting a map of the *E. coli* genome shows that there are *Hfr* strains (PK19, KL96) that have the *F* episome inserted within the bacterial genome adjacent to the *his* operon. These strains could be obtained from bacterial stock centers or the labs that work on them.

    The *F* episome in these strains will occasionally excise to become extrachromosomal again. Though the excision is usually precise, sometimes it is not. Some of the times that the excision is imprecise, flanking DNA including the *his* operon will be included with the excised *F* factor.

    The strategy to identify an *F'* (*his*) is to set up conjugations as follows: *Hfr PK19* (or *KL96*) *ampˢ⁻* × *F⁻ his⁻ ampʳ aziˢ* and select for histidine prototrophs and ampicillin resistance in liquid culture. Then set up conjugations with these *F⁻ his⁺ ampʳ aziˢ* (mostly) or *F' (his) ampʳ aziˢ* (rarely) and a recipient strain, *F⁻ his⁻ ampˢ aziʳ*. Plate on azide and histidine dropout plates. Only the cells that are azide resistant and that have acquired a mobile copy of *his⁺* will survive.

    Final confirmation of the existence of the *F'* can be accomplished by performing conjugation experiments that show there is a high frequency of recombination for the *his* locus but *not* for other loci.

19. a. Lactose, histidine, purines, tryptophan, and phage T1 *should* be added.

    Cysteine, glucose, and galactose *should not* be added.

    b. *cysX⁺ trp⁺ tonᴿ*

c.

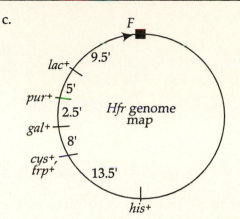

d. Recombination frequency (*trp–cysX*) = recombinants/total where recombinants for *trp* and *cys* are either ++ or – – for the two genes.

Recombination frequency (*trp–cysX*) = 10/200 = 5%.

e. Recombination frequency (*ton–cysX*) = recombinants of selected type/total selected, where recombinants for *ton* and *cys* are either + – or – +. But because *ton* is the selected marker, we cannot identify the – + class. They should be approximately equal in number, however, so we can simply count the *cysX*+ numbers and divide them by only the *ton*^R class (the total selected). By this method, reasonably accurate measures of map distance can be achieved. Recombination frequency (*ton–cysX*) = (10 + 10)/200 = 10%.

f.
```
                    |– 5 mu –|
      — ton —— trp —— cysX —
            |——— 10 mu ———|
```

20. Here is the table again, but now with processed data. The bold markers are the two whose distance is being estimated by the frequency of *cotransduction* (when both marker alleles of the recipient are converted to those of the donor). The less frequent the cotransduction, the farther apart the markers are. In this case, it can be seen that the *aroA pyrD* loci are far apart, being cotransduced only 5% of the time. So those markers should be put at opposite ends of the map with the third marker, *cmlB*, in between. The question that remains is which outer marker is *cmlB* closest to. This is determined by finding which marker has the higher cotransduction frequency, *pyrD*+ or *aroA*. With a 54% cotransduction frequency, it is clear that *pyrD* is closer to *cmlB*. The map is shown below the table.

DONOR	RECIPIENT	SELECTED MARKER	UNSELECTED MARKER*	FREQUENCY
aroA pyrD+	*aroA+ pyrD*	**pyrD+**	*aroA*	0.05
aroA+ cmlB	*aroA cmlB+*	**aroA+**	*cmlB*	0.26
cmlB pyrD+	*cmlB+ pyrD*	**pyrD+**	*cmlB*	0.54

*Ten of the 200 *pyrD*+ colonies in the first transduction were also *aroA*.

```
        ——— aroA ——————— cmlB —— pyrD ———
or, just as correct,
        ——— pyrD —— cmlB ——————— aroA ———
```

15
Gene Structure and Organization

Problems

1. What features of DNA make it especially well suited for being the material responsible for encoding heritable traits in living organisms?

2. In Griffith's classic transformation experiments, he demonstrated the principle of bacterial transformation and postulated the existence of a molecular agent he called "the transforming principle."
   a. What are the *in vivo* and *in vitro* phenotypes of the two strains of *Streptococcus pneumoniae* that Griffith worked with?
   b. What is the gene that is mutated in one of the two strains and is repaired by recombination during the transformation procedure?
   c. What is the outcome of injecting the following combinations into mice?
      a) IIIR + IIIS
      b) IIIR + heat-killed IIIS
      c) IIIR + heat-killed IIIR
   d. Did Griffith prove that DNA is the transforming factor? If so, how?

3. Compare the kind of data that Avery, MacLeod, and McCarty produced in support of DNA as the transforming principle with that produced by Hershey and Chase. Why may the Hershey and Chase experiments have seemed more convincing than Avery's to those who doubted that DNA is the essential genetic material?

4. What two lines of experimentation with tobacco mosaic virus (TMV) illustrated that RNA can in some cases be the primary (and infectious) genetic material?

5. Understanding the chemistry of nucleic acids is becoming increasingly essential as the field of genetics becomes more focused on molecular aspects of genetic questions. On the trinucleotide shown on the following page, label all of the following:
   a. 5′ and 3′ ends
   b. deoxyribose (circle and label)
   c. one phosphate group (circle and label)
   d. bases (circle and label with their specific name)
   e. one phosphodiester bond
   f. the hydrogen atom that differentiates deoxyribose from ribose
   g. the base that identifies this as DNA rather than RNA
   h. the chemical group on the base that defines it as a DNA base
   i. the atoms in each base shown that would participate in base pairing

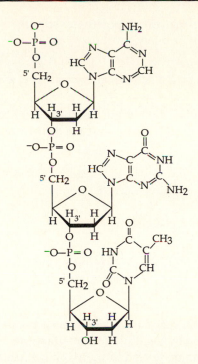

6. DNA usually takes the B-DNA form characterized by right-handed helix direction, with 10 bases per turn of the helix and a very slight (2°) base inclination from the helix axis.

    a. How is Z-DNA different from B-DNA in these respects?

    b. Mark these differences on the illustrations of B-DNA and Z-DNA below.

    c. What support is there for an *in vivo* relevance to a Z-DNA structure for DNA? How might Z-DNA be important for DNA-dependent molecular processes?

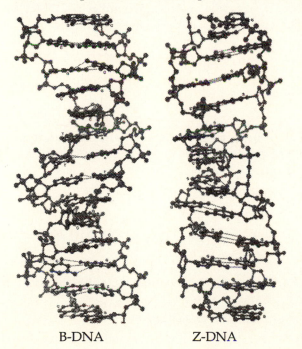

B-DNA   Z-DNA

7. The double helices shown below represent negatively supercoiled, positively supercoiled, and relaxed (but not in that order) B-DNA molecules, each 312 bp long.

   a. Label each structure.

   b. Indicate which topological transitions are catalyzed by DNA topoisomerase I and which are catalyzed by DNA topoisomerase II.

   c. If the supercoiled structures shown were *circular* and had the same number of helical turns as the relaxed structure shown, how many superhelical turns would need to form for the most energetically favored supercoiled forms?

   d. Which molecule would be the least compact *in vivo*?

8. Describe the process of packaging T4 phage DNA by the headful.

9. If the following schematically illustrated bacteriophage T4 genome were to infect an *E. coli,* show how it would replicate, concatamerize, and be packaged by headful. You may assume that the length shown in the drawing below is the length that can be stuffed into one phage head. Label the concatamer and the terminal repeats on your drawing.

A	B	C	D	E	F	A	B

A'	B'	C'	D'	E'	F'	A'	B'

10. Bacteriophage ϕX174 has been a very important model organism in studies of DNA replication. Describe the evidence for each of the following features of the ϕX174 genome:

    a. composed of single-stranded DNA

    b. circular

    c. small (just over 5,000 nucleotides)

    d. overlapping genes

11. Lambda phage has been instrumental in studies of gene regulation and in recombinant DNA technology. One very important feature of lambda phage is its mechanism of replication.

    a. Write the DNA sequence that defines the *cos* sequence.

    b. Show where the *ter* enzymes cut the *cos* sequence.

    c. What is the %GC of the single-stranded complementary ends?

    d. Are circular monomers or linear concatamers of lambda phage DNA packaged into mature lambda phage coats?

12. A total of 78 chromosomes is seen in a karyogram (a display of the metaphase chromosomes of an organism) of a diploid organism.

    a. Is it more likely that the organism is a prokaryote or a eukaryote? Why?

    b. How many unique *types* of chromosomes would there be if similar chromosomes were paired or grouped together?

   c. If the chromosomes were numbered, which would be larger, chromosome 2 or 20?

   d. Briefly describe the G-banding technique and what regions of chromosome are preferentially stained in the procedure.

13. C value is an important measure of genetic complexity. Though it does not perfectly correlate with physiological complexity, C value is generally higher in organisms further diverged from their original ancestor within a taxon.

   a. What is the C value for humans, sea urchins, *Caenorhabditis elegans,* and lily?

   b. Approximately how many base pairs long is a double-stranded DNA molecule that measurements indicate is 1 mm long when fully extended?

   c. How much larger (e.g., 10×, 100×, etc.) is the haploid genome of *Drosophila melanogaster* than that of *Escherichia coli*?

   d. If a single dNMP is 330 Daltons, on average, what is the molecular weight of all the DNA in a single human somatic cell?

14. Histones are an essential class of proteins for packaging DNA into chromosomes.

   a. Approximately what fraction of their amino acids is positively charged at physiological pH? Why is this useful?

   b. Among proteins, histones are some of the best conserved among widely diverged taxonomic groups. Which histone is the exception?

   c. Would you expect there to be more histones per kilobase in euchromatin or in heterochromatin? Why?

15. For each order of chromatin packing that is postulated to bring a eukaryotic DNA molecule's primary nucleotide sequence to the highly compact metaphase chromosome, name the structure characteristic of the compaction level and the process or forces that are required for generating it.

16. The UV_{260} absorbance curve is shown below for fractions of *Drosophila nasutoides* genomic DNA fragments separated by density gradient equilibrium centrifugation in a cesium chloride gradient. GC content for unique sequence DNA averages 40% (fragments of which sediment at about 1.7 g/cm^3).

   a. Label the peak corresponding to the satellite DNA.

   b. In which region would you expect to find the sequence (AATAT)$_n$?

   c. Where would you expect to find a gene encoding alcohol dehydrogenase?

   d. Where would you expect to find centromeric DNA?

   e. Compared to mouse genomic DNA fragments separated by the same procedure, is there anything unusual about the satellite DNA in this example?

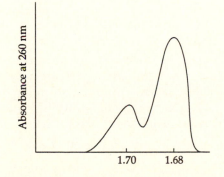

Challenge Problems

17. Assuming Chargaff's rules apply, how often would you expect the stop codons TAG, TAA, and TGA to occur (in any one of six reading frames) in a 3,000 base-pair fragment of double-stranded DNA isolated from an organism with an $(A + T)/(G + C)$ ratio of 1.50?

18. During the process of generating somatic cell hybrids used in the physical mapping of human chromosomes, immortalized rodent cells (which have *unusually* long telomeres) are fused (using cell aggregation–promoting agents such as the sendai virus or polyethylene glycol) with human cells. One curious phenomenon, essential for this chromosome mapping technique, is that most of the human chromosomes are lost over the ensuing cell divisions while all of the rodent chromosomes are retained. In this way, one can obtain cells that by chance contain a subset of human chromosomes, such as X, 7, and 21 only. Using your knowledge of telomerase and telomeres, propose hypotheses to explain the observations that (1) most human chromosomes are lost, (2) a few human chromosomes are retained, and (3) all rodent chromosomes are retained.

Team Problems

19. Prior to Watson and Crick's publication of the structure, DNA was known to be a polynucleotide, to have an approximately equal number of adenines as thymines and guanines as cytosines (Chargaff), and to exist in a highly organized helical structure with two distinctive "regularities" of 0.34 nm and 3.4 nm (Franklin). What bond forces (van der Waals forces, hydrogen bonds, ionic bonds, or covalent bonds) are primarily involved in giving rise to each of the features described above? On a simple ribbon diagram of a DNA double helix, show where each of these forces is acting that leads to the observed structural features.

20. What are three ways to determine that a 100 kb fragment of DNA (of which you have the complete sequence) is human in origin rather than from *E. coli*?

Solutions

1. There are three characteristics expected of genetic material. First, it should be capable of carrying in stable form the information an organism requires for structure, function, development, and reproduction. Second, it should replicate with good accuracy. Third, it should be capable of variation so that organismal change, adaptation, and evolution can occur.

    DNA is a very stable polymer. By having a variable order of bases, and with each base having different chemical properties and sets of bases encoding different amino acids, DNA can encode a nearly infinite variety of RNA and proteins.

    The accuracy of DNA replication is very high because of the inherent chemical properties of the bases and because the enzymes that mediate the process synthesize and edit DNA with high fidelity.

    Finally, DNA is highly capable of variation. Single base changes can change the structure and meaning of an RNA molecule or protein. Each gene, or unit of information, is made up of many (tens to many thousands) of bases, each of which can vary by mutating from A, for example, to C or G or T. Further, the organization of the genes themselves can be changed, adding additional opportunities for variation at the DNA level to lead to variation at the organismal level.

2. a. Griffith worked with R and S strains of *Streptococcus pneumoniae*. S strains grow into smooth, shiny colonies on agar and cause death by pneumonia to mice injected with them. R strains grow into rough, irregular colonies on agar and do not cause death when injected into mice.

   b. The gene mutated in R strains encodes an enzyme necessary for the synthesis of the polysaccharide-rich capsule. The slippery capsule contributes to the smooth and shiny appearance of the bacterial colonies and protects the bacteria bearing it from the host's immune system.

   c. a) rapid death because of the IIIS strain

      b) death because of transformation of the IIIR strain with DNA from the IIIS strain

      c) no death because there is no viable gene present for the enzyme necessary to make the smooth coat

   d. Griffith did *not* show that DNA is the transforming factor. Avery, MacLeod, and McCarty first demonstrated that DNA is the molecule that carries the genetic information necessary for transformation. Hershey and Chase later confirmed the role of DNA beyond doubt.

3. Avery et al. in 1944 showed that purified nucleic acid could transform bacteria and purified protein, lipids, or polysaccharides could not. They went on to show that DNA, not RNA, was capable of transformation by treating their nucleic acid sample with RNase and showing that transformation still occurred, whereas treating nucleic acid with DNase destroyed the transforming material.

   Criticism focused mostly on the incomplete nature of the purification procedure. The possibility that small amounts of critical RNA or protein were associated with the DNA—and in fact were responsible for the transformation of genetic information—could not be excluded.

   Hershey and Chase in 1953 took an entirely different approach. They radio-labeled phage T2 DNA with ^{32}P or radio-labeled phage T2 proteins with ^{35}S. Using these "hot" bacteriophage, they found that phage ghosts (coats minus the DNA) were labeled with the ^{35}S, but little (if any) of what went inside the cell was. None of the T2 phage produced by the infection contained ^{35}S. When Hershey and Chase used ^{32}P to label phage T2 DNA, they found that not only was most of the radio-labeled DNA deposited into the infected cells, but the phage progeny also contained some of the ^{32}P.

   In addition to demonstrating the *presence* of phage DNA (in the form of recycled nucleotides, probably) rather than relying on attempts to demonstrate the *absence* of other components (proteins, RNA), there was something particularly persuasive in the use of radionucleotide labeling rather than simple purifications. Personalities and social contexts may have also played a role, not to mention that Hershey and Chase performed their experiment 9 years after Avery et al., when work on DNA was beginning to have wider interest.

4. First, Gierer and Schramm showed that purified RNA alone could cause lesions like those caused by TMV. Also, RNase treatment completely destroyed the ability of the purified RNA to cause lesions. The second line of experimentation, by Fraenkel-Conrat and Singer, mixed the proteins of one strain of TMV with the RNA of a second strain, and vice versa. In each case, the viruses isolated from the leaf lesions corresponded to those dictated by the RNA component.

5.

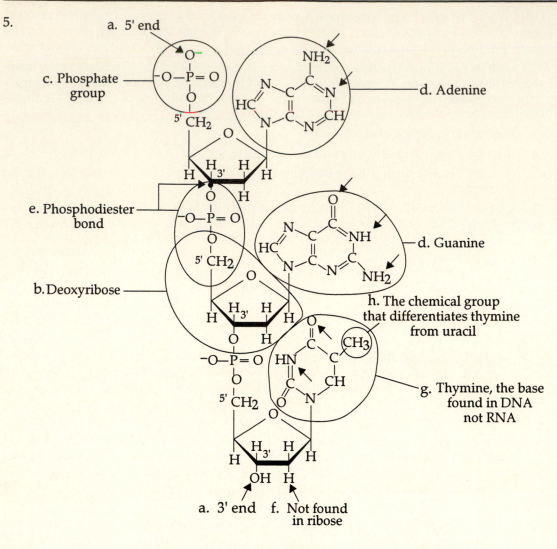

a. 5' end

c. Phosphate group

d. Adenine

e. Phosphodiester bond

d. Guanine

b. Deoxyribose

h. The chemical group that differentiates thymine from uracil

g. Thymine, the base found in DNA not RNA

a. 3' end   f. Not found in ribose

i. Arrows (↖) point to atoms involved in standard Watson-Crick base-pairing.

6. a. Z-DNA is different from B-DNA in that (1) it is a *left-handed* helix, (2) it has 12 bases per turn of the helix, and (3) it has a larger base inclination (8.8°) from the helix axis compared to B-DNA.

b.   Three features that distinguish B-DNA from Z-DNA are illustrated below:

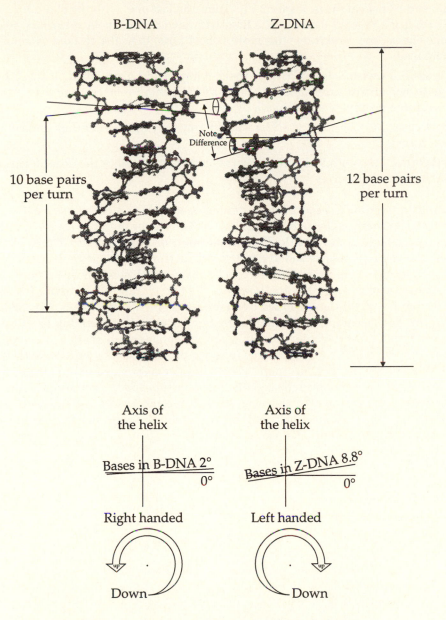

c.   Support for an *in vivo* role for Z-DNA includes the following:

1.   Z-DNA-binding proteins may stabilize Z-DNA structures. DNA binding proteins that recognize Z-DNA have been found in the cell nucleoids/nuclei of organisms as diverse as bacteria, *Drosophila*, and wheat.

2.   Z-DNA may provide a run of left-handed turns that help make local unwinding of right-handed turns easier.

3.   Z-DNA may help promote local DNA unwinding necessary for genetic recombination events.

4.   Z-DNA may be involved in replication or transcription regulation. Some evidence for this exists. Essential SV40 genome replication and transcription control regions contain purine-pyrimidine stretches that readily take on the Z-DNA conformation; proteins that bind and stabilize these regions might promote transcription of adjacent sequences or initiation of DNA replication.

7. a. The number of bases/turn = 312/the number of helical turns = 9.5, 11.6, and 10.4 for the helix with 33, 27, and 30 helical turns, respectively. Thus, the last one, with the standard number of bases/turn, is the relaxed B-DNA. The first, with fewer bases/helical turn, is positively supercoiled. The second structure, with more helical turns than the relaxed form, is negatively supercoiled.

    b. Topoisomerase II untwists relaxed DNA to make it negatively supercoiled. Topoisomerase I adds twists to negatively supercoiled DNA to make it more relaxed, but does not make relaxed DNA more positively supercoiled.

    c. Three in both cases. 30 − 27 = 3 or 30 − 33 = −3 (differing only in whether they are left or right superhelical turns).

    d. The relaxed-form DNA (with 30 helical turns) is least compact. Both the other forms would tend to fold over on themselves.

8. The T4 bacteriophage genome is a *single* linear, double-stranded DNA chromosome that is both circularly permuted and terminally redundant. Both of these latter qualities result in part from two aspects of T4 biology: concatameric DNA replication and the headful DNA packaging mechanism. First, a single double-stranded DNA molecule from the infecting bacteriophage is replicated many times by the *E. coli* replication machinery. Second, some of these short double-stranded molecules denature and reassociate so that the two strands are overlapping only by the ends. Other strands join this lengthening, overlapping chain of T4 genomes. Meanwhile, T4 ligase is sealing the single-stranded molecules together, end-to-end, generating from the short pieces a long, repeating DNA molecule. Later, this long concatamer will be replicated just as the original, shorter genome was. When the time comes for packing DNA into the newly synthesized and assembled phage heads, DNA from this concatamer is threaded in but cannot fit more than what was originally injected into the cell; the phage head capacity is limited. Rather than pulling out the part of the concatamer that was stuffed into the head, the DNA is cut once a "headful" of DNA has been added. What remains of the long concatamer is then brought to another phage head so a headful of DNA can be added to it.

9.

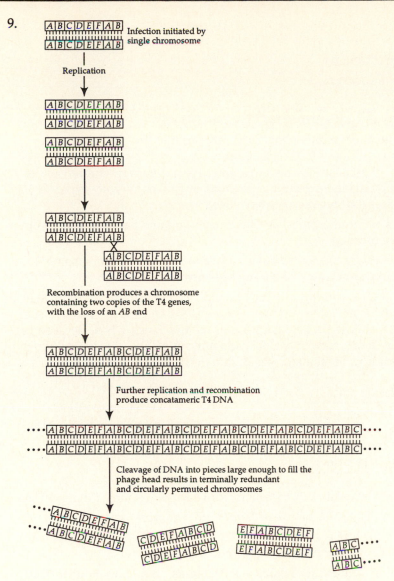

Infection initiated by single chromosome

Replication

Recombination produces a chromosome containing two copies of the T4 genes, with the loss of an *AB* end

Further replication and recombination produce concatameric T4 DNA

Cleavage of DNA into pieces large enough to fill the phage head results in terminally redundant and circularly permuted chromosomes

10. a. Double-stranded DNA obeys Chargaff's rules, which state that the amount of the base adenine will equal the amount of thymine and the amount of cytosine should equal the amount of guanine. The 25% A, 33% T, 24% G, and 18% C base composition of φX174 suggests that the genome must be single-stranded.

b. Both the phage genome's resistance to exonuclease digestion and the appearance of the genome in electron micrographs suggest a circular DNA molecule. Exonucleases are enzymes that can only remove bases one by one from the ends of DNA molecules, but endonucleases cut DNA within the molecule and so could cut both linear and circular molecules.

c. The size of the φX174 genome could be estimated by electron microscopy, centrifugation, or electrophoresis. The exact size can be determined by DNA sequencing.

d. That the genes of φX174 are overlapping was determined after DNA sequencing showed that some of φX174's proteins were encoded by genes that, in a different reading frame, also encoded another φX174 protein.

11. a. and b.

$$\downarrow$$

5′ GTTACGGGGCGGCGACCTCGCGGGT 3′

3′ CAATGCCCCGCCGCTGGAGCGCCCA 5′

$$\uparrow$$

c. %GC = [(G + C) / (A + C + G + T)] × 100 = (10/12) × 100 = 83.3%

d. Neither are. Linear monomers of lambda phage DNA are put into the viral coats.

12. a. The organism is a eukaryote because (1) only eukaryotes have *metaphase* chromosomes, (2) eukaryotes are typically diploid (and prokaryotes are not), and (3) eukaryotes typically have many different chromosomes (but prokaryotes have one).

b. The organism has 39 pairs of chromosomes, probably including one pair that is heteromorphic—the pair of sex chromosomes.

c. By convention, chromosomes are numbered from largest to smallest. Chromosome 2 then will typically be larger than chromosome 20 for any organism.

d. G-banding is a form of Giemsa staining in which metaphase chromosomes are heated or briefly treated with proteolytic enzymes—to open up the DNA and expose reactive sites—before Giemsa staining. G-banding preferentially stains A- and T-rich DNA. Because heterochromatin (such as centromeres, repetitive DNA, and noncoding DNA) is typically rich in adenine and thymine, it tends to be stained more darkly in the G-banding technique.

13. a.

ORGANISM	C VALUES
humans	2.75×10^9 base pairs
sea urchins	8×10^8 base pairs
C. elegans	8×10^7 base pairs
lily	3×10^{11} base pairs

b. Length (in base pairs) is about equal to (2.94 base pairs/nm) × length (in nm) (1 mm = 10^6 nm). Therefore, length (bp in 1 mm) = (2.94 base pairs/nm) × 10^6 nm = 2.94×10^6 base pair

c. The *Drosophila melanogaster* haploid genome is 1.75×10^8 base pairs. The *E. coli* genome is 4.1×10^6 base pairs. Thus, the *Drosophila* genome is about 43 times bigger.

d. Total molecular weight (MW) = number of bases × 330 Daltons/base. The number of bases in a human diploid cell = C × (2 bases/base pair) × (2 [haploid/diploid]) = (2.75×10^9 base pairs/haploid cell) × (2 bases/base pair) × (2 haploid/diploid) = 11×10^9 bases/diploid cell. Therefore, MW (of total cellular DNA) = (330 Daltons/base) × (11×10^9 bases) = 3.63×10^{12} Daltons.

14. a. 0.25 or 25% of the amino acids in histones are lysine or arginine, two positively charged amino acids. Because DNA is negatively charged (the oxygens of the phosphate groups), having positively charged histones of approximately equal abundance allows those charges to be neutralized and compaction of DNA to proceed. More DNA can be contained within a cell and divided in mitosis and meiosis because of histones.

b. Histone H1 contains a highly conserved core sequence but otherwise has lower amino acid identity from organism to organism. There are also multiple H1 histone genes per cell, each encoding slightly different H1 proteins. Some tissues have a different protein altogether; avian red blood cells, for example, have a protein called H5 that serves a function similar to that of H1.

c. Heterochromatin, which is more compact than euchromatin, would be expected to have more histones per kilobase.

15. 1. double-helix: covalent bonds between nucleotides, stacking forces between base pairs, hydrogen bonding between bases, ionic bonds with salt ions to reduce repulsion of negatively charged oxygen atoms.

    2. beads-on-a-string: Histones 2A, 2B, 3, 4 (nucleosome core), and 1 join DNA to make a nucleosome.

    3. 30 nm chromatin fiber: helix of packed nucleosomes

    4. extended section of chromosome: relaxed scaffold proteins attaching to the scaffold attachment region of DNA

    5. condensed section of chromosome: helices of metaphase-chromosome-shaped scaffold bound to DNA to make a metaphase chromosome

16. a. The satellite DNA, strangely enough, is in the larger peak. The smaller peak is the main band, which contains most of the unique sequence = DNA fragments. AT-rich repetitive DNA is less dense than most of the DNA fragments in the main band.

    b. The sequence $(AATAT)_n$ would almost certainly be found in the large, satellite peak.

    c. A gene encoding alcohol dehydrogenase would most likely be found in the main band with the other unique sequences.

    d. Centromeric DNA, highly AT-rich, would be found in the large, satellite peak, to the left of the repetitive sequence in part b.

    e. *Drosophila nasutoides* is extreme in possessing 60% satellite DNA.

17. To solve this problem, you must remember that Chargaff's rules allow you to assume that the amount of deoxyadenosine in a piece of double-stranded DNA will equal the amount of deoxythymidine and the amount of deoxycytodine will equal the amount of deoxyguanosine. This condition arises from the fact that base pairing in dsDNA follows very specific rules of complementarity. Starting from Chargaff's rule, then, we can calculate what fraction of the dsDNA each nucleotide will be:

    $(fA + fT)/(fG + fC) = 1.5$ and $(fA + fT) + (fG + fC) = 1$ where fN = fraction of all bases that is base "N"

    $(fA + fT) = 1.5(fG + fC)$

    $(fA + fT) = 1.5(1 - (fA + fT))$

    $(fA + fT) = 1.5 - 1.5fA - 1.5fT$

    $2.5fA + 2.5fT = 1.5$

    and because $fA = fT$ by Chargaff's rules,

    $5fA = 1.5$

    $fA = 0.3, fT = 0.3$

    and because $f(A + C + G + T) = 1$, $fG + fC = 0.4$; and again, $fG = fC$ so

    $fG = 0.2, fC = 0.2$

    $p(TAG) = 0.3 \times 0.3 \times 0.2 = 0.018$

    $p(TAA) = 0.3 \times 0.3 \times 0.3 = 0.027$

    $p(TGA) = 0.3 \times 0.2 \times 0.3 = \underline{0.018}$

    $p(\text{one of these three}) = \quad 0.063$

    Because we need to find the *number* of these triplets in *one frame* of a strand 3,000 bases long, we multiply by 1,000 (the number of codons in one frame in 3,000 bases).

    $p(TAG \text{ or } TAA \text{ or } TGA)$ in 6,000 bases = $0.063 \times 1,000 = 63$ stop codons predicted to occur by chance in a given reading frame of a 3,000 bp sequence.

18. One possibility (among many) has to do with the unusually long telomeres of the rodent cells used in these experiments or the activity and specificity of the rodent telomerase in comparison to the human telomerase. (1) A hypothesis to explain the loss of human chromosomes, then, is that the cell fusion event is followed by several rounds of cell division during which rodent telomeres are sufficiently maintained or extended (by rodent telomerase), but human chromosome telomeres are not. When the chromosomes encoding human telomerase are lost—or the chromosomes encoding other genes necessary for telomerase expression or activity are lost—loss of human chromosomes would continue toward completion. (2) One hypothesis to explain the retention of a few human chromosomes is that perhaps the retained human chromosomes have had mouse telomeres translocated onto them or have been extended by mouse telomerase (which would need to be a somewhat rare event) so that they are preserved in the hybridoma environment. (3) Finally, a hypothesis to explain the retention of all rodent chromosomes is that their initial length protects them from loss or the rodent telomerase is more active or expressed more abundantly than the human telomerase. Each of these is a disprovable hypothesis.

19. DNA as a polymer, a polynucleotide, arises from covalent phosphodiester bonds between nucleotides along the backbone of DNA molecules.

    Equal amounts of A and T and of C and G arise from the hydrogen bonds between nitrogenous bases in the keto form, holding together the two strands of DNA molecules.

    The helical structure with its characteristic spacing arises primarily from the stacking, or van der Waals (hydrophobic,) interactions between pairs of bases with pairs above and below them.

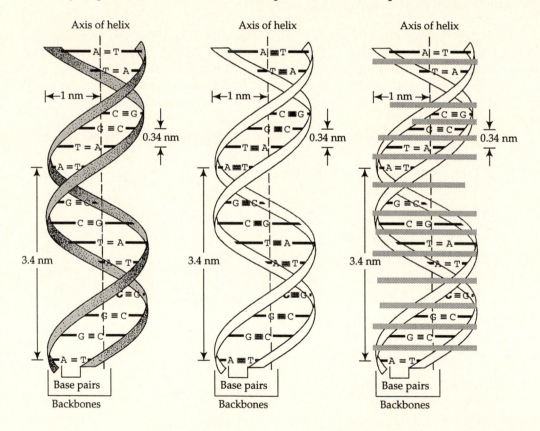

20. (1) Search for Alu I repeats. There should be one about every 5,000 bp in humans and none in *E. coli*. (2) Label the DNA (described in later chapters), heat the DNA to dissociate it, and add it to denatured metaphase chromosomes; if the DNA is from humans, it should find its complementary sequence. If it is bacterial, it won't. (3) Look for long repetitive sequences—As and Ts, ACA-CAC, GCAGCA, and the like. These are much more common in humans than in bacteria. (4) Look for long open reading frames (stretches of DNA that, when read as codons, have many fewer stop codons than one would expect by chance and therefore could encode for the hundreds of amino acids that most protein consists of. Check the sequence of these proteins to known proteins (using a computer and the international gene and protein databases). If any of the sequences match or are very similar to known human proteins, that will be good evidence for the DNA being from a human.

16
DNA Replication and Recombination

Problems

1. In an experiment to investigate the manner in which DNA is replicated, *E. coli* was grown for several generations in minimal medium in which the only nitrogen source was [^{15}N] H_4Cl (ammonium chloride containing an isotope of nitrogen denser than the usual ^{14}N). This process resulted in the bacterial DNA being extensively labeled with "heavy" nitrogen. Sample 0 of the bacterial culture was removed, lysed (cells broken open), and fractionated by equilibrium gradient centrifugation.

    The bacteria remaining were gently sedimented in a low-speed centrifuge and resuspended in medium containing only the "light" ammonium chloride ([^{14}N] H_4Cl) as its nitrogen source. Sample 1 was removed after one cell doubling occurred and processed in the same manner as sample 0. Sample 2 was removed after two cell doublings, and sample 3, after three cell doublings. Both were processed in the same manner as samples 0 and 1.

    a. Which sample contained the greatest proportion of dense DNA?

    b. Which sample contained the least proportion of dense DNA?

    c. Which sample had two bands of approximately equal abundance?

    d. Which samples contained a band that was similar in density to that found in a sample grown exclusively in "light" nitrogen?

    e. Which sample had a single band of a density between that of *E. coli* grown exclusively in "light" nitrogen and that of *E. coli* grown exclusively in "heavy" nitrogen?

2. How would the UV densitometry data from the Meselson-Stahl experiment (described in Problem 1) have been different if DNA replication was conservative? Dispersive?

3. Arthur Kornberg showed that DNA could be synthesized *in vitro* by the addition of DNA fragments, deoxynucleotide triphosphates, and magnesium ions to *E. coli* lysate.

    a. What essential role in DNA synthesis does each component of this *in vitro* assay have?

    b. Draw a balanced chemical reaction illustrating the *in vitro* addition of a single deoxynucleotide triphosphate to a nucleic acid, n deoxynucleotides long [$(dNTP)_n$]. Include all essential elements.

    c. Why did Kornberg include some radiolabeled deoxynucleotide triphosphates in the reaction?

    d. How would knowledge of the nonprotein components of the DNA synthesis reaction be useful for the biochemical purification of a DNA polymerase enzyme?

4. DNA usually takes the form of a double helix of nucleic acid strands in antiparallel orientation, one strand running 5′ to 3′, the other 3′ to 5′. All known DNA polymerases synthesize new DNA strands in the 5′ to 3′ direction. That is, they add deoxynucleotides to the 3′ hydroxyl group (of a primer or existing DNA strand) and extend the polymer of nucleotides in the 3′ direction.

    a. Label 5′ and 3′ ends of the *in vitro* DNA synthesis reaction shown below:

    ————— • • • • • • (direction of synthesis) →
    ————————————————————

    b. When a nucleotide is being added to a growing nucleic acid chain by a DNA polymerase, does the 3′ hydroxyl group or 5′ phosphate of the nucleotide form a bond with the end of the nucleic acid chain?

    c. Addition of ATP is not needed for *in vitro* DNA synthesis. Where does the energy for the polymerization of DNA come from?

    d. How could the 5′ to 3′ directionality of DNA synthesis be demonstrated experimentally using an *in vitro* DNA synthesis reaction?

5. The *E. coli* genome encodes three different DNA polymerases.

    a. Which polymerase has 5′ to 3′ exonuclease (proofreading) activity?

    b. Which polymerase is most abundant in a typical *E. coli*?

    c. Which is also called the "Kornberg enzyme"?

    d. Which has a multiprotein catalytic core in addition to several other subunits?

    e. Which of the three is least well understood?

    f. Which catalyze(s) DNA synthesis in a template-dependent manner?

    g. Which catalyze(s) DNA synthesis in a primer-dependent manner?

6. *In vitro* DNA synthesis of supercoiled DNA requires topoisomerase to reduce the tightness of the DNA twists. Initiator proteins bind to the origin of replication and recruit DNA helicase. DNA helicase denatures the double helix and binds and activates primase.

    a. What monomer does primase polymerize?

    b. Is primase template dependent?

    c. Is primase primer dependent?

    d. What constitutes the primosome?

    e. How are leading and lagging strands different in their requirement for primase-synthesized primers during DNA synthesis?

7. Hundreds of SSB proteins are found at (and near) each replication fork during DNA synthesis in *E. coli*.

    a. What are some of the important functions of these SSB proteins?

    b. Would you expect more SSB proteins to be bound to the template of the replisome's leading strand or lagging strand?

8. Consider a replication fork that moves (opens) at 500 nucleotides/second.

    a. How fast does the helix ahead of the replication fork need to rotate to prevent the formation of excessively supercoiled DNA? (Hint: there are 10 base pairs per turn of the double helix.)

    b. What enzyme plays the major role in preventing excess twisting?

9. Eukaryotic DNA is replicated in a semidiscontinuous manner.

   a. What does it mean for DNA replication to be discontinuous?

   b. What about *semi*discontinuous?

   c. Illustrate your answer with a labeled diagram of a replication bubble.

10. Several viral genomes (e.g., that of lambda phage) and episomal DNAs (e.g., F factor) replicate according to the rolling circle model.

    a. Is rolling circle replication bidirectional?

    b. Do DNAs replicating by rolling circle form theta structures?

    c. Is rolling circle replication semiconservative?

    d. What occurs at the origin of replication in DNA that replicates by rolling circle that isn't required when DNA is synthesized according to the replication bubble model?

    e. Is the F factor DNA transferred through the conjugation pilus as a newly synthesized single-stranded molecule, an "old" single-stranded molecule, or a double-stranded molecule that is half new and half "old"?

11. The genome of *Drosophila* replicates in about 3 minutes. The genome of *E. coli* replicates in about 20 minutes. *Drosophila* have about 100 times more DNA than *E. coli*. DNA polymerase progression in *Drosophila* is also about 20 times slower than in *E. coli*. How can you explain *Drosophila's* faster replication of a much bigger genome?

12. Autonomously replicating sequences (ARSs) that can serve as origins of replication for DNA molecules in yeast cells have been found in the yeast *Saccharomyces cerevisiae*.

    a. What are the molecular elements that make up an ARS?

    b. Which of these elements appear to be required?

    c. What is the function of the nonobligatory elements?

    d. Which elements are bound by the origin replication complex (ORC)?

    e. How does the recruitment of ARS-binding factor 1 (Abf1) to the ARS affect replication?

13. Telomerase, an enzyme with both RNA and protein components, lengthens the telomeres at the end of eukaryotic chromosomes.

    a. Why is there a need for a special enzyme to assist with the replication of chromosome ends?

    b. Does the telomerase enzyme use an RNA or DNA template when it adds nucleotides to a DNA strand?

    c. Does telomerase add ribonucleotides or deoxyribonucleotides?

    d. Does the telomerase enzyme add nucleotides to the underhang or overhang of the chromosome end?

    e. Which enzymes complete the extension of the 5' end of the DNA molecule?

14. Unlike most protein-encoding genes, histone genes are transcriptionally most active from the end of G_1 to just before the end of S phase.

    a. Why would you expect histone genes to be transcribed while most other genes are quiescent during DNA synthesis?

    b. How does the cell stop histone production as DNA synthesis nears completion?

15. Propose an experiment that would test whether nucleosome histones are segregated to daugher chromosomes conservatively (i.e., old histones remain together in their original nucleosome, and newly synthesized histones are assembled in new nucleosomes) or semiconservatively (nucleosomes of daughter chromosomes contain mixtures of old and new histones). How would you assess whether newly synthesized and "old" histones mix and match to give nucleosomes composed of both old and new? How would you distinguish this model of histone replication from one where "old" nucleosomes retain their particular histone composition while new nucleosomes are made from newly synthesized histones?

16. UV and X-ray irradiation increase not only the mutation rate in cells, but also the frequency of crossing-over. How does this observation support the idea that single-strand breaks initiate the recombination process?

17. The *E. coli polAI* gene encodes DNA polymerase I. Why do mutations in this gene that cause a loss of 5′ to 3′ DNA synthesis activity (yet retain the exonuclease activity) result in increased mutation rate when the bacteria are exposed to UV light or chemical mutagens?

18. One of the compelling aspects of the Holliday model of recombination is that it provides an explanation for the mechanism of gene conversion.
    a. How can base mismatches arise from heteroduplex formation?
    b. What enzymes are involved in the gene conversion process in yeast?
    c. What is the result of mismatch repair enzymes not being able to distinguish between the "old" and the "new" DNA strand by the time the cell reaches prophase I of meiosis?
    d. For an organism of genotype m^+/m, can an m^+ allele be converted to m just as often as to m^+?

19. a. What is gene co-conversion?
    b. Illustrate co-conversion of two alleles, c and d, in an organism that has the genotype:

$$\frac{\bullet a^+\ c\ \ d\ \ b^+}{\bullet a\ \ c^+\ d^+\ b}$$

Challenge Problems

20. Progression through the cell cycle is a carefully controlled process in eukaryotes.
    a. What are the phases of the cell cycle?
    b. In what order does the cell progress through these phases?
    c. What are the main events that occur in each phase?
    d. Where does cytokinesis typically occur?
    e. What are the functions of cell cycle checkpoints?
    f. Under what cellular conditions does a cell arrest at the G_2 checkpoint?
    g. How can we explain the fact that the M and G_1 checkpoints occur slightly before the end of their respective phases instead of exactly in between the phases?
    h. What is the important consequence of the proteolytic degradation of the cyclins after they have bound to Cdk at START and at the G_2 checkpoint?

21. The two cell types known to have detectable telomerase activity are germ-line cells and tumor cells.

    a. What do both of these cells have in common?

    b. What evidence is there that telomere length is regulated genetically?

    c. What effect would you expect a potent telomerase inhibitor to have on a patient's fast-growing tumor?

    d. Telomerase knockout mice that are homozygous for a nonfunctional telomerase gene have been produced. Several generations of these mice seem perfectly normal. Eventually, however, a sterile generation of descendants is born. Explain.

Team Problems

22. Though there are many models to explain recombination, the Holliday model provides the foundation for most current thinking on the recombination process.

    a. What are the stages of the Holliday model of recombination?

    b. What is a strand invasion?

    c. What is a Holliday intermediate?

    d. What is branch migration?

    e. What is the difference between heteroduplexes, patched duplexes, and spliced duplexes?

    f. What feature of the Holliday model allows one to predict that a physical exchange between two gene loci will result in genetic exchange of flanking chromosome markers only about 50% of the time?

    g. What is the role of DNA synthesis in the recombination process according to the Holliday model?

23. A small section of the duplex DNA belonging to one chromatid of each of the two homologous chromosomes is shown below. Arrows indicate the location of the single-strand breaks that *initiate* the process of recombination. The asterisks show where the second set of single-strand breaks that lead to the resolution of the Holliday intermediate occurs. Note: **bold** base pairs are different between the two chromatids.

```
3' — TTT GGG CCC AGA TTT GGG CCC AAA  →
5' — AAA CCC GGG TCT AAA CCC GGG TTT  →  centromere
                ↑                 *


                ↓                 *
5' — AAA CCC GGG TTT AAA CCC GGG TTT  →
3' — TTT GGG CCC AAA TTT GGG CCC AAA  →  centromere
```

    a. Draw a Holliday recombination intermediate.

    b. Draw the resolved chromosomes.

    c. If in this particular case, the mismatched purine-pyrimidine pairs are corrected by the removal of the purine-containing nucleotide, circle the letter representing the nucleotide that would be changed, and indicate what nucleotide it would be replaced by.

    d. Does the exchange of genetic material shown here lead to a recombination of flanking markers?

Solutions

1. The experiment described in this problem is very similar to the classic Meselson-Stahl experiment that demonstrated that DNA replicates by a semiconservative mechanism. To solve the problem, remember that both strands of the bacterial chromosomal DNA contain the denser isotope when sample 0 is taken; hence, the duplex DNA is "heavy." When the nitrogen source is changed from $[^{15}N] H_4Cl$ to $[^{14}N] H_4Cl$, the nucleotides that synthesized *de novo* are less dense; duplex DNA made from these less dense nucleotides will be less dense in proportion to how much they contain the "light" nitrogen.

   After one round of cell division in media containing "light" nitrogen, the newly synthesized strands will be light. That is, in sample 1, for every old "heavy" strand, there will be one new "light" strand. Because the equilibrium density centrifugation does not denature duplex DNA into single strands, this half "light," half "heavy" DNA will sediment to a layer midway between the "heavy" and "light" duplex DNA.

   After a second round of DNA replication in "light" nitrogen, newly synthesized strands will again be "light." But this time half of the old strands are "light" and half are "heavy." Because DNA is replicated semiconservatively, half the chromosomes will be *all* light, and half will be half "heavy" and half "light."

   After a third round of DNA replication in "light" nitrogen, even more of the chromosomes will be entirely light, which will be detected as an even thicker band (or higher peak in the densitometry) at the position in the centrifuge tube corresponding to entirely "light" DNA.

   a. sample 0

   b. sample 3

   c. sample 2

   d. samples 2 and 3

   e. sample 1

2. The following data represent densitometry scans of duplex DNA banded in centrifuge tubes by density equilibrium gradient centrifugation. The position of the peak on the x axis indicates the average density of the DNA in the band. Peak height corresponds to proportion of DNA in the sample that is of specified density.

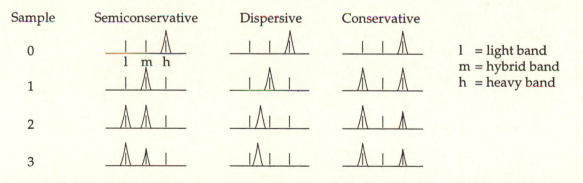

3. a. The DNA fragments serve as templates and perhaps as primers for DNA polymerase. Deoxynucleotides triphosphates serve as monomers that are polymerized by DNA polymerase and as carriers of high-energy bonds that drive the DNA synthesis reaction. Magnesium ion is an essential cofactor for DNA polymerase activity. *E. coli* lysate contains DNA polymerase.

    b.
    $$[(dNTP)_n] + dNTP \xrightarrow{\text{DNA polymerase, } Mg^{++}} [(dNTP)_{n+1}] + PP_i$$

    c. The radio-labeled nucleotides were included so that newly polymerized nucleic acids could be distinguished from the preexisting nucleic acid (template).

    d. Defining the essential nonprotein components necessary in an *in vitro* reaction—and having a way to determine the amount of DNA synthesized—allowed Kornberg's group to assay (determine) the amount of DNA polymerase "activity" in different fractions of *E. coli* lysate. They could separate large proteins from small and negatively charged from positively charged and neutral proteins, for example, and test each of the fractions for activity. By following the fractions with increased (more concentrated) activity, they were able to purify the protein away from most other contaminating ones.

4. a. 5′ ———•••••• 3′ (direction of synthesis) →
        3′ ——————————————————— 5′

    b. All DNA polymerases add deoxynucleotides such that the 5′ phosphate of the nucleotide is bound to the 3′ hydroxyl group of the growing nucleic acid chain.

    c. The energy necessary for nucleic acid synthesis comes primarily from the dNTPs being added to the growing chain. The high-energy bond between the alpha and beta phosphate is broken to allow the formation of the bond between nucleotide and nucleic acid chain.

    d. There are many ways to demonstrate 5′ to 3′ directionality of DNA polymerases using an *in vitro* DNA synthesis reaction. One would be to add a primer designed to the *3′ end* of a single-stranded DNA template and quantitate the product synthesized. Another primer, designed to the *5′ end* of the same single-stranded DNA template, could be used in the *in vitro* DNA synthesis reaction. Radio-labeled dNTPs will be incorporated only when the primer is made to the 3′ end of the template as shown below:
        5′ —primer—•••*•••*•• 3′   (primer extended)   5′ —primer— 3′ (no extension)
        3′ ——template—— 5′   3′ ——template—— 5′

5. a. DNA polymerase I
    b. DNA polymerase I
    c. DNA polymerase I
    d. DNA polymerase III
    e. DNA polymerase II
    f. all DNA polymerases
    g. all DNA polymerases

6. a. primase polymerizes ribonucleotides (for the most part)
    b. Yes, primase is template dependent.
    c. Primase needs no primer; it frequently makes the primer for DNA polymerase.
    d. The primosome is made up of DNA (at the replication fork), helicase, and primase.
    e. Both leading and lagging strand synthesis is initiated by primase, but the lagging strand needs frequent repriming because of the discontinuous nature of its synthesis.

7. a. Single-strand binding proteins (SSBs) have many functions. They include contributing to the opening and unwinding of duplex DNA at the origin, sustaining unwinding of duplex DNA by helicase at replication forks, improving the fidelity of and elongation by *E. coli* DNA polymerase II and III, and protecting single-stranded DNA from degradation by nucleases.

    b. More SSBs would be expected to bind the template of the lagging strand because that strand must remain single stranded until enough duplex DNA has been denatured for priming by primase and DNA synthesis by DNA polymerase III. The leading strand has no such lag period.

8. a. $$x \text{ rotations/second} = \frac{500 \text{ base pairs/second}}{10 \text{ base pairs/helical turn}} = 50 \text{ rotations/second}$$

    b. Topoisomerase II (gyrase) is the major *E. coli* enzyme that removes excess supercoils caused by unwinding of the double helix at the replication fork.

9. a. Discontinuous replication refers to the repeated priming and DNA synthesis events required for synthesis of the lagging strand. These RNA-DNA molecules are called Okazaki fragments. They are eventually converted into a completely deoxyribonucleic acid by the removal of the ribonucleotides in the primers (by DNA polymerase I), the replacement of those excised ribonucleotides (also by DNA polymerase I), and the ligation of adjacent DNA fragments (by DNA ligase).

    b. Eukaryotic DNA is replicated in a *semi*discontinuous manner because only one of the two strands of DNA at the replication fork is synthesized discontinuously. The leading strand is synthesized in a continuous manner.

    c.

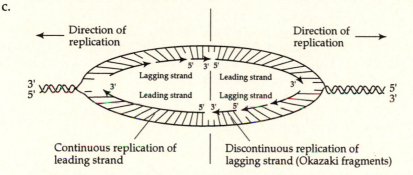

10. a. Rolling circle DNA replication is bidirectional.

    b. DNA replication by the rolling circle mechanism does *not* form theta structures.

    c. Rolling circle DNA replication is semiconservative.

    d. Rolling circle DNA replication requires a single-strand nick that is not necessary in the replication bubble mechanism.

    e. The *F* factor DNA transferred through the conjugation pilus is an "old" single-stranded molecule. It serves as a template for replication after it arrives in the recipient cell.

11. *Drosophila* can replicate its entire genome more quickly than *E. coli* because the *Drosophila* genome has approximately 3,500 origins of replication and the *E. coli* circular chromosome has only one.

12. a. ARSs can contain at least four different sequence elements, A, B1, B2, and B3.

  b. The A element appears to be required by ARSs.

  c. The nonobligatory B elements affect the frequency with which the origin is used.

  d. The A element and B1 and/or B2 form the core of the ARS that is bound by the origin recognition complex (ORC).

  e. Binding of ARS-binding factor (Abf1) to the B3 element enhances replication from the origin containing the B3 element.

13. a. A special enzyme is needed to replicate chromosome ends because DNA polymerases are unable to synthesize new nucleic acid without a primer in the 3′ to 5′ direction. As such, there will always be a 5′ underhang that is unable to be filled in by DNA polymerase without assistance.

  b. Telomerase adds nucleotides using an RNA template.

  c. Telomerase adds deoxyribonucleotides.

  d. Telomerase adds nucleotides to the 3′ overhang.

  e. Primase adds an RNA primer. DNA polymerase I or III extends the primer. Ligase bonds the newly synthesized DNA chain to the adjacent DNA chain. DNA polymerase I removes the ribonucleotides of the primer.

14. a. Sufficient numbers of histones must be synthesized in order to bind the entire length of the newly synthesized chromosomes in S phase.

  b. Histone transcription slows near the end of DNA synthesis, and the histone mRNAs are degraded, preventing further translation.

15. One approach would be to grow the eukaryotic cells in isotopically labeled amino acids for several generations, change the growth medium to include only nonlabeled amino acids, and then allow one round of DNA replication and cell division to occur. Although most of the old histones would be labeled, the new histones would be largely unlabeled. If one could isolate the nucleosomes of the daughter chromosomes without them disassociating, one could assess whether the nucleosomes were of intermediate density (i.e., made up of both heavy and light histones) or were a mixture of heavy nucleosomes (i.e., made up entirely of heavy histones) and light nucleosomes (i.e., made up entirely of light histones).

16. UV and X-ray irradiation both cause breaks in DNA strands. If recombination is initiated by single-strand breaks, it is possible that radiation-induced breaks could substitute for action of the endogenous enzyme and that the rest of the steps in the recombination process would proceed normally.

17. *E. coli* polymerase I is involved in removal of RNA primers, replacement of those ribonucleotide bases with deoxyribonucleotide bases, and DNA repair processes. Because the mutations described in this problem do not destroy the 5′ to 3′ exonuclease activity, the DNA polymerase I enzyme can still remove the ribonucleotides as is required in chromosome replication. The DNA polymerase III enzyme is capable of filling in the gaps that are produced by the removal of primers. But DNA polymerase III is generally not active in DNA repair and so cannot make up for the loss of DNA polymerase I in this respect. More mutations will remain unrepaired and be transmitted to daughter cells.

18. a. Base mismatches arise from heteroduplex formation because homologous chromosomes can contain many polymorphisms in their sequences. If one chromosome has an A at a given position and the homologous chromosome has a G at that same position, if the 5' to 3' oriented strands of each chromosome switch partners (as during strand invasion), a C-A mispair and a T-G mispair will form.

    b. Gene conversion requires a DNA polymerase and DNA ligase, in addition to the other enzymes necessary for recombination.

    c. Without the ability to distinguish the old strand from the new strand, the cell cannot correct mismatches to the base pairing that existed in the older duplex. Therefore, in the process of gene conversion, deleterious mutations should be corrected no more frequently than normal alleles will be mutated.

    d. yes, for the same reasons described above

19. a. Gene co-conversion is when more than one allele gets converted to another type by gene conversion. Co-conversion usually occurs for genes that are very near each other.

    b.

$$\frac{\quad\quad\quad \bullet a^+\ c\ \ d\ \ b^+}{\quad\quad\quad \bullet a\ \ c^+ d^+ b} \quad \longrightarrow \quad \frac{\quad\quad\quad \bullet a^+\ c^+ d^+ b^+}{\quad\quad\quad \bullet a\ \ c^+ d^+ b}$$

20. a. gap phase 1, synthesis phase, gap phase 2, mitosis

    b. Cells progress through the phases of the cell cycle as follows: $G_1 \rightarrow S \rightarrow G_2 \rightarrow M$.

    c. G_1: most synthesis of nonreplication-related proteins, specialized cell functions

    S: DNA synthesis, histone synthesis

    G_2: synthesis of proteins necessary for chromosome condensation and mitosis

    M: sister chromatids segregated to opposite poles of the cell, preparations for cytokinesis

    d. Cytokinesis, if it occurs, follows mitosis.

    e. Cell cycle checkpoints provide opportunities for the cell to assess its readiness to proceed to the next phase. Cells may stall, arrest, or activate a cell death program, depending on the situation.

    f. Cells will arrest at the G_2 checkpoint if all of the cell's DNA is not replicated, the cell isn't big enough, or the environment is not favorable.

    g. M and G_1 checkpoints occur slightly before the end of their respective phases because the cell is not completely poised to enter the next phase; further preparations are still required. Mitosis can begin immediately upon satisfaction of the requirements of the G_2 checkpoint.

    h. Degradation of the cyclins after they have bound to (and have activated) Cdk allows the transition from active to inactive form of Cdk and the silencing of the signal that initiates the transition to a new phase.

21. a. Both cell types are capable of unlimited cell division as long as they are supplied with necessary nutrients.

    b. Evidence for genetic control of telomere length includes the existence of mutations in *E. coli* genes TEL1 and TEL2 that cause cells to maintain their telomeres, but at a consistently shorter length.

    c. A telomerase inhibitor might be expected to restore cell division limits to the tumor cells, causing their progeny eventually to senesce and die. The effectiveness of the therapy would depend on how long the telomeres were at the beginning of the treatment and how many cell divisions are left.

d. It is thought that the knockout mice are viable and sterile for several generations because of the excess length of mouse telomeres. They can be shortened in the germ line for several generations and still not encroach on essential genes. Beyond a critical number of generations, however, the germ-line telomeres can shorten no more without damaging some of those essential genes.

22. a. The stages of the Holliday model of recombination are:
   1. recognition and alignment of two homologous DNA double helices
   2. nicks made in the same strand of the two homologous double helices
   3. strand invasion of the opposite double helix and base pairing with the invaded helix
   4. ligation of the invading strand to the broken end of the reciprocal invading strand, producing the Holliday intermediate
   5. branch migration that generates complementary regions of hybrid DNA in both double helices
   6. Cleavage by endonuclease and ligation of the gaps to produce either patched duplexes or spliced duplexes

   b. Strand invasion is when a strand of one double helix leaves its complementary binding partner and forms base-pair interactions with the strand of the homologous double helix that is complementary to it.

   c. A Holliday intermediate is the fully ligated stage of strand-invaded recombination intermediate structure. While the four strands of DNA have the structure of a Holliday intermediate, branch migration can occur. The Holliday intermediate is resolved to the recombination product by a second set of cleavage and religation reactions.

   d. Branch migration is the translocation of the crossed strands invading (in either direction) along the length of the two chromatids involved in the recombination event.

   e. Heteroduplex is a general term that describes the complementary hydrogen bonding between two strands of DNA that are not completely complementary or are not usually paired together. Patched duplexes are (parental) chromosomes that have a short stretch, or patch, of a single strand of DNA from another (homologous) chromosome. Spliced duplexes are (recombinant) chromosomes that have a short stretch of a single strand of DNA from another (homologous) chromosome and have recombined such that the duplex DNA on one side of the patch is from one chromosome and the duplex DNA on the other side of the patch is from the homologous chromosome.

   f. Because the *orientation* of the last cleavage and religation reaction in the resolution of the Holliday intermediate is predicted to be random, one should see patched duplexes (parental type chromosomes) and spliced duplexes (recombinant chromosomes) with equal frequency. Therefore, 50% of the time the physical exchange will lead to recombination, and 50% not.

   g. DNA synthesis is not a necessary component of the Holliday model, but is for some variants of it. DNA synthesis may occur at the strand invasion stage prior to branch migration. DNA synthesis is, of course, expected to occur during the repair of base mismatches that might arise when patches of heteroduplex DNA are formed.

23. a.

```
       3' — TTT GGG CCC AGA TTT GGG CCC AAA  →  centromere of chromosome 1
       5' — AAA CC:                GGG TTT  →
                  ↑      heteroduplex DNA    *
    site of first ligation        ← branch
                         heteroduplex DNA
                                             *
       5' — AAA CC:                GGG TTT  →  centromere of chromosome 1'
       3' — TTT GGG CCC AAA TTT GGG CCC AAA  →
```

b.

```
       3' — TTT GGG   CCC AGA TTT GGG   CCC AAA  →  centromere of chromosome 1
       5' — AAA CC:C GGG TTT AAA CCC :GGG TTT  →
                   heteroduplex DNA        ↑
                                    site of second ligation
                   heteroduplex DNA        ↓
       5' — AAA CC:C GGG TCT AAA CCC :GGG TTT  →  centromere of chromosome 1'
       3' — TTT GGG   CCC AAA TTT GGG   CCC AAA  →
```

c. The bold G would be changed to an A in the upper chromosome. The bold A would be changed to a G in the lower chromosome.

d. In this case, the flanking markers are not recombined (i.e., they are in the parental configuration), and the only evidence of the genetic exchange are the patched duplexes in the two chromosomes that were involved.

Transcription, RNA Molecules, and RNA Processing

Problems

1. Transcription is the RNA polymerase–mediated process that generates an RNA molecule (or transcript) using a DNA template.

   a. What are the molecular requirements of a transcription reaction? What does each component contribute to the reaction?

   b. Compare and contrast the enzymology of transcription and DNA replication.

2. RNA polymerase catalyzes the growth of an RNA molecule by adding ribonucleotides to the 3′ end of another ribonucleotide or a chain of ribonucleotides.

   a. Does RNA polymerase begin transcribing the DNA template strand from the gene template's 5′ or 3′ end? Illustrate with a diagram.

   b. What molecule most directly contributes the bond energy for ribonucleotide polymerization?

3. The sequence below represents the first 20 bases of a primary RNA transcript from an *E. coli* gene. Draw the duplex DNA that encoded it. Label template and nontemplate strands and 5′ and 3′ ends.

         5′ AUCGGACCAUUCGCGUCUGG ... 3′

4. Although the bacteria examined to date have only one RNA polymerase, eukaryotic organisms have three.

   a. Which RNA polymerase synthesizes each of the following types of RNA: mRNA, tRNA, rRNA, and snRNA?

   b. Which RNA polymerase is found in the nucleolus? What is the main biosynthetic process that occurs in the nucleolus?

   c. If you wanted to study how long a eukaryotic cell can live without the ability to transcribe protein-encoding genes, what compound might you treat cells with? Why?

5. Termination of transcription in *E. coli* can occur by two main mechanisms: a *rho*-dependent pathway and a *rho*-independent pathway.

   a. What could be the role of the sequence with twofold symmetry and the AT-rich string of bases that follows it for the functioning of the *rho*-independent pathway?

   b. Give two differences between the two transcription termination pathways.

6. Unlike eukaryotes, prokaryotes are characterized by coupled transcription and translation, complete colinearity of gene and mature transcript, and the production of unprocessed transcripts. How do eukaryotes differ from prokaryotes in these respects? How are they similar?

7. The removal of introns from primary transcripts in eukaryotes is a carefully regulated process, often restoring reading frames that had been disrupted by the introns.
   a. Are introns removed from the DNA molecule?
   b. Are introns transcribed? Translated? Explain.
   c. How do R-looping experiments demonstrate the presence of introns within genes?

8. The spliceosome is made up of small nuclear ribonucleoprotein particles (snRNPs).
   a. What are snRNPs made up of?
   b. Which of the six principal snRNPs binds first to the intron? What does it bind to?
   c. To what part of the intron does the U2 snRNP bind?
   d. What secondary structure is formed with the assistance of the first two snRNPs to bind to the intron?

9. Molecules of rRNA constitute an essential component of ribosomes in both bacteria and eukaryotes.
   a. What is the function of ribosomes?
   b. Describe their general structure.
   c. What RNA polymerase transcribes rRNAs in eukaryotes? Where?
   d. What might be the function of rRNA within the ribosome?

10. *E. coli* ribosomes, like mammalian ribosomes, are made of both rRNA molecules and proteins.
    a. Are there more proteins or RNA molecules in a single *E. coli* ribosome?
    b. Is a greater fraction of an *E. coli* ribosome's molecular weight due to proteins or RNA?
    c. How many more rRNAs are used to construct a mammalian ribosome than a ribosome in *E. coli*?

11. Not all rRNA genes in eukaryotes are part of the rDNA transcription unit.
    a. Which eukaryotic rRNA gene is *not* transcribed by RNA polymerase I?
    b. Which eukaryotic RNA polymerase transcribes the rRNA gene in part a?
    c. Which eukaryotic rRNA gene has an internal promoter?

12. Unlike most pre-rRNAs, the pre-rRNA of the protozoa *Tetrahymena* contains introns that must be removed during RNA processing.
    a. Where are the *Tetrahymena* rRNA introns removed?
    b. What is particularly unusual about these splicing reactions?
    c. How is the removal of the spacers between rRNAs dramatically different from the removal of the intron within the rRNA exons?

13. How is the removal of group I introns different from the removal of introns from mRNAs?

14. What is the role of TFIIIA, TFIIIB, and TFIIIC in the transcription of 5S rRNA- and tRNA-encoding genes?

15. Describe the process of intron removal from the yeast pre-tRNA.Tyr.

16. Pre-tRNAs are extensively processed to generate the mature charged tRNA.

    a. List five of those modifications.

    b. What is the function of each of those modifications?

    c. Label the anticodon, loops I–IV, a stem, and the location of amino acid attachment on the secondary structure of yeast alanine tRNA shown below.

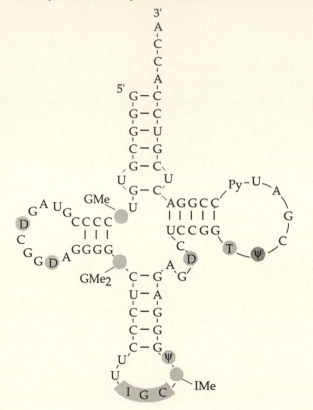

17. What are the functions of RNase P and RNase Q in the processing of tRNAs in *E. coli*?

Challenge Problems

18. The following duplex DNA sequence is part of the sequence near the beginning of a transcribed region in the *E. coli* genome. Bases are numbered in the standard manner.

```
    -40         -30         -20         -10         +1          +10
     •           •           •           •           •           •
5′ ATTCTTGACATTTTTCATAAAATTTGGTATAATACATAACATCGATAGGA  ...
3′ TAAGAACTGTAAAAAGTATTTTAAACCATATTATGTATTGTAGCTATCCT  ...
```

    a. Underline and label the bases that constitute the consensus sequence for *E. coli* promoters. Label the Pribnow box.

    b. Where will the RNA polymerase holoenzyme bind first?

    c. Circle the region containing the base pairs of duplex DNA that is unwound by RNA polymerase during the early steps of transcription initiation.

    d. What is the sequence of RNA that will be transcribed from this sequence (based on the way the sequence is numbered)?

e. If the –10 region were changed from 5' TATAAT 3' to 5' GATAAT 3', what might happen to the rate of transcription?

f. If this promoter were one recognized by sigma factor 54 (σ54) what bases in the promoter might you expect to be different?

19. The nontemplate strand of a fragment of a eukaryotic protein-encoding gene is shown below. Exon-intron boundaries are marked by a short vertical line. The numbers above the DNA sequence correspond to the number of the base in the sequence shown. The DNA sequence continues at the end of the line on the next line of DNA sequence below.

A "conceptual translation" of this piece of DNA is also shown in three frames. Briefly, amino acids that correspond to each of three possible reading frames the DNA sequence could have are shown (in the single letter code form). If the first reading frame is the one used, then the codon "TTA" encodes for the first amino shown, "L" the amino acid leucine. If the second reading frame was used, then "TAG" would have coded for a translational stop, indicated by a period, and that would have been the last amino acid in the protein. If the third reading frame was used, then "AGT" would have coded for "S," the amino acid serine. The underlined amino acids are the *actual* amino acids encoded by the exons in this example.

```
exon                             27                                54

TTA GTT ATT TGC GGG ACC AAT TCG TAC AAA CCC CTC TGT CGG ACG TAC GCA TTT
 L   V   I   C   G   T   N   S   Y   K   P   L   C   R   T   Y   A   F
  .   L   F   A   G   P   I   R   T   N   P   S   V   G   R   T   H   L
   S   Y   L   R   D   Q   F   V   Q   T   P   L   S   D   V   R   I   .


   · intron begins                81                               108

AAG|GTA GGT CCA GTC GTG CTG ACG CAA TCG TCA TTT GCA CTC CAC ATA ATT CTA
 K   V   G   P   V   G   L   T   Q   S   S   F   T   L   H   I   I   L
  R   .   V   Q   S   V   .   R   N   R   H   L   H   S   T   .   F   .
   G   R   S   S   R   F   D   A   I   V   I   Y   T   P   H   N   S   R


   · exon 2 begins               135                               162

G|GA GGG AAA GTA CCT GGT TGA GAA AGA AGT AGA AGG GAT AGG CTT GTG TCC ATA
 G   G   K   V   P   G   .   E   R   S   R   R   D   R   L   V   S   I
  E   G   K   Y   L   V   E   K   E   V   E   G   I   G   L   C   P   Y
   R   E   S   T   W   L   R   K   K   .   K   G   .   A   C   V   H   T
```

a. Above the DNA sequence (nontemplate strand) or on another piece of paper, write in the unprocessed mRNA sequence that would be transcribed from this region of the gene.

b. Find and underline the branch point sequence in the unprocessed mRNA.

c. Draw a line connecting the guanine from the 5' end of the intron to the adenine of the branch point.

d. Draw a line connecting the last base of the first exon to the first base of the second exon.

e. Is a codon in the final, processed mRNA interrupted by an intron in the genomic DNA?

Team Problems

20. Sucrose gradient centrifugation is a useful technique for separating RNA molecules, riboprotein complexes, and organelles based on their size and degree of compaction. Imagine a procedure in which mammalian cell extracts are loaded onto a sucrose gradient and spun at high speeds for several hours. Fractions are then taken by puncturing the bottom of the centrifuge tube and collecting fractions in numbered tubes.

    a. How would you detect the presence of RNA in each fraction?

    b. Which RNA species (rRNA, mRNA, or tRNA) would you expect to be most abundant among the RNA species collected?

    c. In what order would you expect to collect fractions containing the following rRNAs: 18S, 28S, 5.8S, small ribosomal subunit, large ribosomal subunit?

    d. If an RNA molecule with a significantly greater molecular weight sediments in the same band as an RNA with a lower molecular weight, what would you predict about the degree of the compaction of the lower molecular weight RNA?

21. RNA processing is important in both prokaryotic and eukaryotic rRNA production.

    a. Sketch the structure of the rDNA transcription unit in both prokaryotes and eukaryotes.

    b. Describe the rDNA promoter structure.

    c. Which sequence-specific DNA-binding proteins bind the promoter of the rRNA transcription unit in eukaryotes?

    d. List the processing steps that are required to produce mature rRNAs in both groups.

    e. What enzyme is known to catalyze some of these processing steps in prokaryotes?

Solutions

1. a. A transcription reaction requires a DNA molecule to serve as template for transcription with a promoter (and, *in vivo*, transcription factors) to indicate where to begin transcribing and which strand to transcribe. Transcription reactions also require an RNA polymerase that recognizes the promoter on the DNA, NTPs to serve as the building blocks and donors of bonding energy, and a buffered salt solution for optimal RNA polymerase activity.

    b. DNA polymerases polymerize **deoxyribonucleotides;** RNA polymerase polymerizes ribonucleotides. DNA polymerases require a **primer** to provide a 3′ OH group to add deoxyribonucleotides to; RNA polymerase needs no such primer. DNA polymerases generally begin DNA replication at **origins of replication;** RNA polymerases begin transcribing at the promoters of genes.

        Both DNA and RNA polymerases synthesize chains of nucleotides in a template-dependent fashion. Both polymerases add nucleotides by connecting the (5′) alpha-phosphate of a nucleotide triphosphate to the 3′ OH of the last base that was added (i.e., in the 5′ to 3′ direction) and interact with other proteins that assist in their functioning.

2. a. RNA polymerase begins transcribing a DNA template strand from the 3′ end of the template.

        ```
        3′ ...    TACTGCTA ... 5′    DNA template
        5′        AUGACGAU ... 3′    Transcript
        ```

    b. Ribonucleotides contribute most of the bond energy for ribonucleotide polymerization. The release of the pyrophosphate (PP_i) accompanies the formation of the phosphodiester bond.

3.   5'   `AUCGGACCAUUCGCGUCUGG` ... 3' Transcript

                5' ...   `ATCGGACCATTCGCGTCTGG` ... 3' Nontemplate strand

                3' ...   `TAGCCTGGTAAGCGCAGACC` ... 5' Template strand

4. a. In eukaryotes, RNA polymerase I synthesizes all rRNA except the 5S rRNA. RNA polymerase II synthesizes mRNA and some snRNAs. RNA polymerase III synthesizes tRNA, some snRNAs, and the 5S rRNA.

     b. RNA polymerase I is found in the nucleolus, where it contributes most of the rRNAs necessary for **ribosome synthesis**, the main biosynthetic process that occurs in the nucleolus.

     c. Alpha-amanitin, made by the death-cap toadstool *Amanita phalloides*, allows one to study how long a eukaryotic cell can live without the ability to transcribe genes because it primarily inhibits the action of RNA polymerase II, the enzyme responsible for transcribing protein-encoding genes.

5. a. The *in vivo* function of the sequence with twofold symmetry and the AT-rich string of bases that follows it have not been definitively characterized. It is likely, however, that the formation of a hairpin loop causes the RNA polymerase to pause. The hairpin may fit poorly in the transcript groove of the RNA polymerase and weaken the binding of the RNA polymerase to the transcription bubble. The weak hydrogen bonding of the AT-rich region immediately following the stretch of twofold symmetry may further contribute to the destabilization of RNA polymerase binding.

     b. *Rho*-dependent transcription termination requires additional protein factors (the hexameric protein, *rho*, that binds to the *rut* site on the RNA), ATP hydrolysis, and *no* stretch of uracil residues at the 3' end of the RNA.

        Rho-independent transcription termination is more direct. It requires a GC-rich region of twofold symmetry followed by an AT-rich string of bases that encode for polyuracil in the transcript.

        These events occur in both pathways. (1) The synthesis of RNA stops, (2) the RNA molecule is released from the DNA, and (3) the RNA polymerase is released from the DNA molecule.

6. Transcription and translation are coupled in prokaryotes because, without a nuclear membrane, ribosomes can bind to nascent mRNA as soon as the 5' end of the mRNA has been synthesized well before transcription is over. In eukaryotes, transcription and post-transcriptional nuclear processing and targeting must be completed before cytoplasmic translation can begin.

    Genes in eukaryotes usually contain introns. For example, base 500 of a gene is not necessarily the same as base 500 in the mature transcript, even if both are numbered from the same first base. If there were no introns, eukaryotic genes would also be colinear.

    Eukaryotic transcripts are processed in three ways: (1) Eukaryotic pre-mRNAs are capped at their 5' end (by the addition of a guanine nucleotide, usually 7-methyl guanosine, via a 5'–5' linkage). The first two nucleotides of the transcript are also usually modified by the addition of two methyl groups to each. (2) Most eukaryotic pre-mRNAs are modified on their 3' end by the addition of approximately 50–250 adenine nucleotides. These adenines are added at a sequence called the poly(A) site, just downstream from a consensus polyadenylation signal (AAUAAA). The poly(A) tail is important for conferring stability on the mRNA. (3) Most eukaryotic pre-mRNAs and, much more rarely, rRNAs and tRNAs, contain introns that are spliced out. Pre-mRNA introns are removed by spliceosomes. Pre-rRNA introns are removed by a different form of spliceosome or in some cases have been shown to catalyze their own removal. The rare cases of tRNA splicing are mediated by a different pathway.

Both eukaryotic and prokaryotic RNAs are processed in the following ways: (1) Pre-rRNAs and pre-tRNAs are modified by cleavage of primary transcripts that contain more than one gene product. These RNAs may then be further modified, in eukaryotes, by removal of internal and external spacer sequences and, in prokaryotes, by removal of leader, spacer, and trailer sequences. (2) Pre-tRNAs are also chemically modified to alter some of their bases to such modified bases as pseudouridine. (3) The sequence 5'-CAA-3' is added to the 3' end of the pre-tRNA.

7. a. No, introns are not removed from the DNA molecule. Organized DNA rearrangements that occur in some specific instances (e.g., IgG gene rearrangement) are not RNA splicing. Splicing occurs in primary transcripts, mostly in pre-mRNA molecules.

   b. Introns are transcribed, but are removed (by splicing, in the nucleus) before the RNA is exported from the nucleus and can be translated. Mutations that destroy exon-intron boundaries often produce truncated or aberrant proteins. Introns typically contain stop codons. If not, they may have a number of bases that is not a multiple of 3 and shift the frame of the transcript. Even if the frame is maintained, the intron will encode for amino acids that, if inserted into a protein, may reduce its function.

   c. R-looping experiments reveal the presence of introns within genes by giving evidence to the incomplete colinearity of the gene and transcript. Intronic sequences in genomic DNA will have no sequence to hybridize with in mature mRNA and will therefore loop out as single-stranded DNA. The exonic sequence in DNA and RNA can join in complementary base-pair interactions to form double-stranded hybrids.

8. a. snRNPs consist of proteins and small nuclear RNAs.
   b. U1 snRNP is first to bind to the intron, binding at the 5' end of the intron.
   c. U2 snRNP binds next—to the branch point.
   d. The first two snRNPs, U1 and U2, to bind to the intron form a loop. These recruit other snRNAs.

9. a. Ribosomes serve as scaffolds for protein synthesis. They are RNA-protein complexes that catalyze the formation of peptide bonds between amino acids. Ribosome-based protein synthesis is mRNA-dependent. They also require charged tRNAs as the adapter molecules that align the amino acid to be added.

   b. Ribosomes consist of a small and a large subunit. Each of these subunits has at least one RNA molecule and many small proteins. When the two subunits have loaded onto the mRNA, the mRNA runs through a groove between them.

   c. Three of the four rRNAs (5.8S, 18S, and 28S) are transcribed by RNA polymerase I in the nucleolus of eukaryotes. The fourth rRNA (5S) is transcribed by RNA polymerase III.

   d. rRNA could be having a catalytic role in the ribosome, helping identify the Shine-Dalgarno sequence, for example.

10. a. There are more proteins (54) than RNA molecules (3) in a single *E. coli* ribosome.
    b. A greater fraction of an *E. coli* ribosome's molecular weight is due to RNA. RNA accounts for 2/3 of the *E. coli* ribosome's molecular weight.
    c. A mammalian ribosome has about 85 different proteins, 31 more than an *E. coli* ribosome. Mammalian ribosomes have 4 RNAs, 1 more than *E. coli*.

11. a.   The eukaryotic 5S rRNA gene is not part of the rDNA transcription unit and is not transcribed by RNA polymerase I.

   b.   The 5S rRNA gene is transcribed by RNA polymerase III.

   c.   The 5S rRNA gene has an internal promoter.

12. a.   The *Tetrahymena* rRNA introns are removed in the nucleolus during the processing of the pre-rRNA.

   b.   These *Tetrahymena* rRNA splicing reactions are unusual in that they are self-splicing, requiring no proteins to catalyze the process.

   c.   The removal of the spacers between rRNAs' *cleavage* reactions results in cut RNA that is not religated. The removal of the intron that lies between the rRNA exons involves both cleavage reactions and the subsequent joining of the rRNA exons.

13. The removal of group I introns in *Tetrahymena* rRNA splicing is different from mRNA splicing in several ways. First, group I introns are self-splicing, requiring no protein catalysts. Second, group I introns, cleaved at the 5' end, don't join to the branch point, as in mRNA splicing, but rather have a guanine nucleotide added to their 5' ends. Third, the joining of the exons occurs such that the linear intron is liberated. The intron then circularizes to form a lariat and is cleaved to form linear and circular pieces.

14. TFIIIA binds to *box C* in the 5S rRNA gene internal control region (ICR), allowing TFIIIC to bind to the *box A* region. TFIIIB then binds to the other transcription factors, but not to the DNA directly. TFIIIB then helps position RNA polymerase III 50 base pairs upstream from the beginning of *box A*, near the +1 position of the gene.

15. Introns are removed from the yeast pre-tRNA.Tyr by cleavage by a specific endonuclease and rejoining of the exons by RNA ligase.

16. a.   tRNAs are (1) processed by cleavage from other tRNAs in the same transcript, (2) trimmed to remove sequence at 5' and 3' ends, (3) appended at the 3' end with the sequence 5' CCA 3', (4) spliced (in rare cases) to remove intronic sequence, (5) chemically modified to convert standard bases to bases such as pseudouridine, inosine, and dihydrouridine, and (6) charged by addition of an amino acid to the 3' end.

   b.   Cleavage and trimming remove excess bases to allow proper secondary structure to form. Intron removal also allows proper secondary structure to form. Addition of CCA allows charging of the tRNA with an amino acid by amino acyl synthetases. Chemical modification allows proper secondary and tertiary structure to form and recognition by charging enzymes and interaction with the ribosome to occur. Amino acid addition allows the tRNA to contribute to protein synthesis at the ribosome.

c.

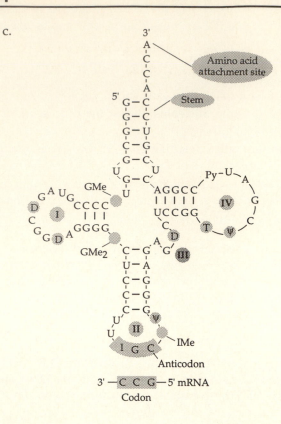

17. RNase P catalyzes the removal of the 5′ leader sequences of *E. coli* pre-tRNAs. RNase Q catalyzes the removal of the 3′ trailer sequences of the same pre-tRNAs.

18.

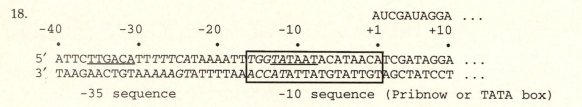

```
                                                    AUCGAUAGGA ...
   -40        -30        -20        -10        +1        +10
    •          •          •          •          •          •
5′ ATTCTTGACATTTTTCATAAAATTTGGTATAATACATAACATCGATAGGA ...
3′ TAAGAACTGTAAAAAGTATTTTAAACCATATTATGTATTGTAGCTATCCT ...
       -35 sequence                  -10 sequence (Pribnow or TATA box)
```

a. −10 sequence (Pribnow, TATA box) and −35 sequence are labeled and underlined.

b. The RNA polymerase holoenzyme binds to the −35 sequence first.

c. The first 17 base pairs to be untwisted are boxed.

d. The mRNA transcript is shown above the +1 base of the gene.

e. If the −10 region were changed from 5′ TATAAT 3′ to 5′ GATAAT 3′, one could expect the rate of transcription to be *reduced* because RNA polymerase would bind less efficiently. It is a deviation from the −10 consensus sequence.

f. If the promoter shown above were one recognized by sigma factor 54 (σ^{54}), one would expect that the TTTCA at −26 (italicized) would be GTGGC (or close to it) and the TGGTA at −14 (italicized) would be TTGCA (or, again, close to that sequence). σ^{54} binds to upstream elements of *E. coli* genes that are expressed when the amount of usable nitrogen in the local environment is low.

```
19.    top line = mRNA                27                              54
       UUA GUU AUU UGC GGG ACC AAU UCG UAC AAA CCC CUC UGU CGG ACG UAC GCA UUU
       TTA GTT ATT TGC GGG ACC AAT TCG TAC AAA CCC CTC TGT CGG ACG TAC GCA TTT
        L   V   I   C   G   T   N   S   Y   K   P   L   C   R   T   Y   A   F
         .   L   F   A   G   P   I   R   T   N   P   S   V   G   R   T   H   L
          S   Y   L   R   D   Q   F   V   Q   T   P   L   S   D   V   R   I   .


                                       81                             108
       AAG GTA GGU CCA GUC GUG CUG ACG CAA UCG UCA UUU GCA CUC CAC AUA AUU CUA
       AAG|GTA GGT CCA GTC GTG CTG ACG CAA TCG TCA TTT GCA CTC CAC ATA ATT CTA
        K   V   G   P   V   G   L   T   Q   S   S   F   T   L   H   I   I   L
         R   .   V   Q   S   V   .   R   N   R   H   L   H   S   T   .   F   .
          G   R   S   S   R   F   D   A   I   V   I   Y   T   P   H   N   S   R


                                      135                            162
       G GA CCC AAA GUA CCU GGU UGA GAA AGA AGU AGA AGG GAU AGG CUU GUG UCC AUA
       G|GA GGG AAA GTA CCT GGT TGA GAA AGA AGT AGA AGG GAT AGG CTT GTG TCC ATA
        G   G   K   V   P   G   .   E   R   S   R   R   D   R   L   V   S   I
          E   G   K   Y   L   V   E   K   E   V   E   G   I   G   L   C   P   Y
            R   E   S   T   W   L   R   K   K   .   K   G   .   A   C   V   H   T
```

a. The unprocessed mRNA sequence that would be transcribed from this region of the gene is written in above the DNA sequence (the nontemplate strand).

b. The branch point sequence, UGCUGAC in this case, is underlined. Note it is also between 18 and 38 bases from the 3′ end of the intron.

c. A solid line (——) connects the guanine from the 5′ end of the intron to the adenine of the branch point.

d. A dashed line (- - -) connects the last base of the first exon to the first base of the second exon.

e. In this particular case, no codon is interrupted by the intron. The last codon of exon 1 is AAG; the first codon of exon 2 is GAG. However, the reading frame is shifted relative to what it would have been without the intron.

20. a. RNA can be detected and quantitated in each fraction by ultraviolet absorbence spectroscopy. RNA's maximal absorption is of UV light with a wavelength of 260 nm.

b. In most organisms, rRNA would be the most abundant species of RNA (approximately 80% in mammals), tRNA, the next most (about 15%), and mRNA, the least abundant (only about 3% of the total).

c. One would expect to collect fractions containing organelles and RNAs in the following order: first (near the bottom of the tube) large ribosomal subunit, then small ribosomal subunit, then 28S, 18S, and 5.8S, in that order.

d. If an RNA molecule with a significantly greater molecular weight sediments in the sucrose gradient to the same band as an RNA with a lower molecular weight, the lower molecular weight RNA must be *more compact* than the higher molecular weight RNA. Normally, sedimentation is a balance between hydrodynamic qualities (such as compactness, which increases density).

21. a.

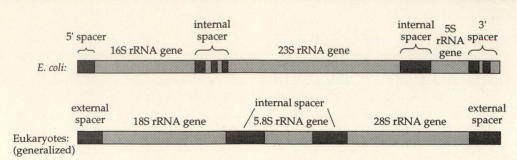

   b. RNA polymerase I transcribes only the rDNA repeat unit. The promoter of the rDNA repeat units has two domains: a core promoter element (+7 to –45) and an upstream control element (UCE at –107 to –186).

   c. The two promoter elements create binding sites for transcription proteins hUBF and SL1. These transcription factors form a transcription complex that promotes both RNA polymerase I binding and transcription by RNA polymerase I. hUBF binds both promoter elements and activates transcription. SL1 consists of the TATA binding protein (TBP) and three TBP-associated factors (TAFs) and is required for RNA polymerase I to recognize the promoter and to initiate transcription.

   d. In prokaryotes, all three rRNAs are transcribed together to make a single precursor rRNA. The transcript is bound by ribosomal proteins and is cleaved into three pre-rRNAs by RNase III cutting within internal spacer regions. Each pre-rRNA is trimmed still further by secondary processing enzymes to generate mature rRNAs.

   In eukaryotes, rRNA is transcribed from the rDNA repeat unit. The transcript is then cleaved by site-specific ribonucleases to remove the ITS and ETS sequences. The 32S pre-rRNA undergoes folding that allows intramolecular base pairs to form. The cleavage of 32S pre-rRNA at the "hinge" region generates the 28S and 5.8S rRNAs in hydrogen-bonded configuration.

   The 5S rRNA is transcribed by RNA polymerase III, producing a mature 5S rRNA without further removal of sequence.

   e. RNase III catalyzes the cleavage of the *E. coli* primary transcript into three pre-rRNAs.

18

The Genetic Code and Translation

Problems

1. The primary structure of a protein is the amino acids sequence of the polypeptide chain(s).
   a. What are the secondary, tertiary, and quaternary levels of protein structure?
   b. Describe the different intra- and intermolecular bonds that are employed at each level of protein structure.
   c. At which levels can a single mutation affect protein structure?

2. Synthesis of bacterial proteins begins with the placement of a tRNA bearing a formyl-methionine (fMet) at the translation initiation codon, AUG. Formylated peptides act as potent attractants to many white blood cells, the cells responsible for the immune response to bacterial infection.
   a. Draw the complete chemical structure of the bacterial tripeptide N-formyl-methionyl-leucinyl-phenylalanine (fMet-Leu-Phe); label the N and C termini.
   b. Circle the peptide bonds in this peptide.
   c. Describe the chemical nature of each of the R groups in this peptide.

3. The determination that amino acids are encoded by trinucleotide sequences, or "codons," was an important step toward deciphering the genetic code.
   a. How does the fact that most proteins are constructed from 20 different amino acids lead one to predict at least a three-base code?
   b. How does a triplet code encoding 20 amino acids produce a degenerate genetic code?
   c. Describe how the *rII* mutant reversion experiments done by Francis Crick's research group helped establish the three-base code.

4. If RNA trinucleotides of the sequence 5′ UGG 3′ were radio-labeled and mixed with ribosomes and charged tRNA molecules, what amino acid would be found on those tRNAs that bound to the radio-labeled fragment?

5. Descibe one exception to the generally true idea that the genetic code is universal.

6. List all the tRNA anticodons (5′ to 3′) that could recognize the codon for:
   a. alanine
   b. cysteine
   c. leucine

7. The specificity for codon recognition lies in the tRNA molecule and not in the amino acid it carries.
   a. How was the role of the tRNA molecule in determining specificity demonstrated?
   b. How would mRNA translation be affected if a tRNA.Tyr were made with an anticodon 5′ IUA 3′ (a change that still allows recognition by the tRNA.Tyr aminoacyl-tRNA synthetase)?
   c. How would mRNA translation be affected if a tyrosinyl-tRNA synthetase were mutated such that it added tyrosine to tRNAs with the anticodon 5′ GAA 3′?

8. The amino acid that initiates a new peptide chain is brought to the ribosome by a tRNA dedicated to that task.

   a. What is the proper designation of the charged tRNA in prokaryotes and in eukaryotes that carries that initial amino acid?

   b. What amino acid is added to the translation-initiating tRNA molecule in prokaryotes?

   c. What enzyme is responsible for formylating methionine?

   d. Does the same aminoacyl synthetase charge the tRNA that recognizes the AUG codon, whether or not the AUG codon is initial or internal?

9. In prokaryotes, the mRNA region that binds the 3' end of the 16S rRNA is called the Shine-Dalgarno sequence.

   a. What is the sequence of nucleotides in the 16S rRNA of *E. coli* that binds to the Shine-Dalgarno sequence?

   b. What ribosomal subunit is the 16S rRNA part of?

   c. What change in translation rate would be expected if the Shine-Dalgarno sequence was mutated so that the 16S rRNA forms more hydrogen bond interactions with it?

10. Protein synthesis in prokaryotes is initiated by the binding of a 30S ribosomal subunit and several initiation factors to an mRNA.

    a. Which other components of the initiation complex must be recruited before the 50S subunit will bind?

    b. Which site (A or P) does the next charged tRNA after the fMet-tRNA.fMet enter?

    c. What molecule provides the energy required for the formation of the initiation complex?

11. Elongation of the polypeptide chains during translocation is a multistep, enzyme-catalyzed process requiring an input of energy.

    a. What are the steps required to prepare a charged tRNA for entry into the A site of the 70S ribosome?

    b. What are the steps of the polypeptide elongation cycle?

    c. Which enzyme catalyzes the formation of peptide bonds?

    d. What degradative enzyme can completely eliminate the activity of the enzyme in part c?

12. Formation of a new peptide bond and removal of the spent tRNA are closely linked processes in peptide synthesis.

    a. What is the E site in the 50S subunit of the ribosome?

    b. What step just *precedes* the translocation of the ribosome?

    c. What common small molecule is produced during the formation of a peptide bond?

13. In both prokaryotes and eukaryotes, many ribosomes translate simultaneously from the same mRNA, allowing more protein to be made more quickly from the same mRNA molecule. On the diagram of a polysome shown below, label the following:

    a. 5' and 3' ends

    b. the stop codon

    c. N- and C-termini of the protein(s)

d. the first ribosome (of those shown) to bind the mRNA

14. Describe the distinct roles of the three *E. coli* release factors, RF1, RF2, and RF3. Are all three required to complete the synthesis of one particular peptide?

15. Many secreted or membrane-associated proteins are processed extensively before being shuttled to the cell envelope. What are the functions of the following?
    a. signal recognition peptide
    b. signal peptidase
    c. glycosyl transferases

16. Proteins and mRNAs are targeted to cellular compartments by specific sequences in the peptide chain or transcript.
    a. What are some of the differences among transit sequences, signal sequences, and nuclear localization sequences?
    b. Unlike other targeting signals, the nuclear localization sequence is *retained* after the arrival of the protein in the nucleus. Why is retention of the nuclear localization sequence important?

Challenge Problems

17. The generation of an aminoacyl-AMP molecule requires ATP and an aminoacyl-tRNA synthetase specific for the tRNA being charged. The same enzyme transfers the amino acid to the appropriate tRNA.
    a. Draw the chemical structure of cysteine attached to the 3′ carbon of the adenine nucleotide of a tRNA.Cys.
    b. Which of the amino acid's chemical groups, carboxyl or amino, will be attached first to the polypeptide chain?

18. One of the important exceptions to the rule that the genetic code is unambiguous is the case of the UGA codon that can sometimes be read not as "stop," but as a codon encoding an amino acid. Describe this unusual exception, how it is effected, and how it is regulated.

Team Problems

19. The assignment of amino acids to particular codons was done by several techniques, including analysis of proteins synthesized using cell-free translation of simple RNA polymers, RNA copolymers of random sequence, and RNA copolymers of known sequence.
    a. What codons would be found in RNA copolymers consisting of 25% uracil and 75% cytosine in random sequence?
    b. What amino acids would be found in polypeptides synthesized in a cell-free system using the RNA copolymer described in part a?
    c. What would be the approximate percentage of each of the amino acids in part b?

d. What percentage of each amino acid would one expect if polypeptides were synthesized in a cell-free system using an RNA copolymer of 75% uracil and 25% cytosine in random sequence?

e. What amino acids would be expected (and in what ratios) if the copolymer translated were poly(UC), in which uracil and cytosine alternate precisely?

20. What does it mean that the genetic code:

a. is continuous ("comma-free")?

b. contains start and stop signals?

c. is nonoverlapping?

d. is degenerate?

e. is universal?

f. is nonambiguous?

Solutions

1. a. The secondary level of protein structure includes the folding and twisting of polypeptide chains into a variety of shapes, such as alpha-helices and beta-sheets. The tertiary level of protein structure is the three-dimensional structure into which the secondary structure is folded. At this level, amino acids that are distantly positioned in the primary structure can be brought into proximity. Finally, the quaternary level of protein structure refers to how multiple polypeptide chains are packaged into multimeric complexes. Not every polypeptide is part of a multimer, and therefore not every peptide has a quaternary level of structure.

b. Peptide bonds generate the primary level of protein structure. Weak electric and hydrostatic bonds between NH and CO groups of amino acids near each other in the polypeptide chain generate the secondary level of structure.

The tertiary level of structure is formed by weak electric and hydrostatic bonds, interaction of hydrophobic groups, and covalent bonds, such as those between cysteines, called disulfide bonds. The tertiary-level structure typically is a result of interactions between parts of the polypeptide that are not adjacent to each other in their primary sequence, but are brought near by folding and the interaction of secondary structures.

The quaternary structure is generated by a similar complement of weak bonds as those that generate tertiary structure. Often the large number of weak bond interactions taken together gives rise to a very strong interaction between the polypeptide chains in a multimeric complex. Disulfide bonds can also exist between polypeptides that make up the quaternary structure.

c. A single base mutation can change an amino acid (altering the primary structure), prevent the formation of an alpha helix (an element of the secondary stucture), change the folding and conformation of the polypeptide (the tertiary structure), and thereby prevent heterodimerization (an example of tertiary structure).

2. a., b.

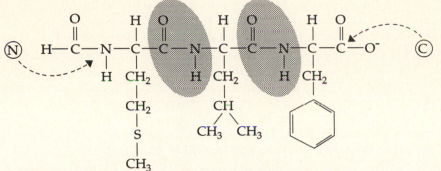

c. All R groups in fMet-Leu-Phe are *neutral, nonpolar.*

3. a. If each codon were only one base long, then the four bases of mRNA (adenine, cytosine, guanine, uracil) could make up only four codons, which would be only four amino acids. If each codon were two bases long, then there would be the capacity to encode 4^2 or 16 amino acids. This is because for each of the four bases at the first position in the codon, there are four bases that could be at the second position. If each codon were three bases long, then the coding capacity expands to 4^3 or 64 amino acids. A three-base codon is therefore sufficient for the needs of organisms with 20 amino acids.

    b. Because all codons encode something—an amino acid or a stop—the 64 codons of the triplet code have excess coding capacity. Most amino acids are encoded by several codons; hence, the genetic code is degenerate.

    c. The *rII* mutant reversion experiments of Francis Crick and his colleagues helped establish the three-base code by showing that an *rII* mutant containing a total of three closely clustered one-base insertions restored the reading frame and could encode a functional protein. Similarly, combining three closely clustered one-base deletions could also revert the mutant phenotype. Again, three one-base deletions would be expected to restore the reading frame if a three-base codon exists.

4. Tryptophan would be found on the charged tRNAs that bound the radio-labeled 5' UGG 3'. Only the anticodon of Trp-tRNA.Trp (5' CCA 3') can base-pair with that fragment.

5. AUA encodes isoleucine in cytoplasmically translated genes, but encodes methionine in the mitochondria. This is an example of context-dependent "ambiguity" of the code—though within one context or the other, the translation isn't ambiguous. In yeast mitochondria, CUN encodes for threonine, and in human mitochondria, CUN encodes for leucine.

6. a. GCU, GCC, GCA, GCG, NGC (where N = A, C, G, U, or I)

    b. UGU, UGC, ACA, GCA

    c. UUA, UUG, CUU, CUC, CUA, CUG, CAA, UAA, NAG (where N = A, C, G, U, or I)

7. a. To show that the mRNA codon recognizes the tRNA and not the amino acid carried by the tRNA, a cysteine residue was attached to the tRNA.Cys *in vitro*. The attached cysteine was chemically converted to alanine, and this Ala-tRNA.Cys was used in the *in vitro* synthesis of hemoglobin. Hemoglobin, synthesized *in vivo*, contains one cysteine each in its alpha and beta chains. However, when the hemoglobin synthesized was analyzed, both chains contained alanine in place of cysteine.

b. If a tRNA.Tyr were made with an anticodon 5′ IUA 3′ and were still recognized by the tRNA.Tyr aminoacyl-tRNA synthetase, many proteins translated in the cell would have been longer because the tRNA would allow tyrosines to be added to a protein at codons UAA that usually indicate stop. I, inosine, recognizes not only U and C, but A as well.

c. If a tyrosinyl-tRNA synthetase were mutated such that it added tyrosine to tRNAs with the anticodon 5′ GAA 3′, tyrosines would be added to proteins sometimes when the codon UUC called for phenylalanine. Though these amino acids are similar in many ways, tyrosines can perform specialized functions (such as serving as sites for phosphorylation).

8. a. The proper designation of the charged tRNA that carries the amino acid that initiates a translation is fMet-tRNA.fMet in prokaryotes and Met-tRNA.Met in eukaryotes.

b. Methionine is added to the translation-initiating tRNA.fMet molecule in prokaryotes to make Met-tRNA.fMet and then formylated to make fMet-tRNA.fMet.

c. The enzyme in *E. coli* responsible for formylating methionine is called *transformylase*.

d. Yes, in *E. coli*, the same aminoacyl synthetase charges the tRNAs that recognize the AUG codon whether the AUG codon is initial or internal.

9. a. In *E. coli*, the sequence of nucleotides at the 3′ end of the 16S rRNA that binds to the Shine-Dalgarno sequence is 5′ ACCUCCUU 3′.

b. The 16S rRNA is part of of the 30S ribosomal subunit.

c. If the binding by 16S rRNA is significantly increased by mutating the sequence upstream from an AUG on transcript, one would expect that the rate of translation initiation would also increase.

10. a. Before the 50S subunit will bind, the 30S initiation complex in *E. coli* must consist of an mRNA, a 30S ribosomal subunit, initiation factors IF1, IF2, and IF3, GTP, and fMet-tRNA.fMet.

b. The next charged tRNA after the fMet-tRNA.fMet enters the ribosome enters the A site. The fMet-tRNA.fMet is in the P site.

c. One GTP molecule provides the energy required for the formation of the initiation complex.

11. a. First elongation factor EF-Tu-GDP binds EF-Ts, releasing GDP. Then, EF-Tu-Ts binds GTP and releases EF-Ts. Finally, EF-Tu-GTP binds aminoacyl-tRNA.

b. First aminoacyl-tRNA enters the ribosomal A site. Then a peptide bond forms between adjacent amino acids catalyzed by peptidyl transferase. Next ribosomes translocate, facilitated by EF-G and hydrolysis of GTP. Finally, the uncharged tRNA ejects from the E site to open up the A site again.

c. Peptidyl transferase catalyzes the formation of peptide bonds.

d. Ribonuclease T1 can completely destroy peptidyl transferase activity, suggesting that at least a major component of peptidyl transferase is RNA.

12. a. The E site in the 50S subunit of the ribosome is the site the spent, uncharged tRNA is sent to after the peptide bond has formed, liberating it from the peptide chain. From the E site, the tRNA *exits* the ribosome.

b. Just prior to ribosomal translocation, the peptide bond is formed.

c. The formation of a peptide bond generates a molecule of water.

13.

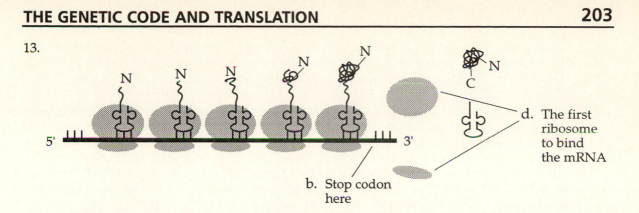

5'   3'

d. The first ribosome to bind the mRNA

b. Stop codon here

14. RF1 recognizes codons UAA and UAG. RF2 recognizes UAA and UGA. RF3 stimulates the termination events rather than recognizing any of the stop codons. Binding of release factors in the A site induces peptidyl transferase to hydrolyze the bond between the polypeptide and the tRNA in the P site. Note, however, that it is *not* necessary to have all release factors for a given termination event. Eukaryotes require only one release factor.

15. a. The *signal recognition particle* (SRP) is a complex of the small 7S RNA and six proteins. The SRP binds to the signal sequence at the N-terminus of the protein, blocks further translation, and escorts the nascent polypeptide-SRP-ribosome-mRNA complex to a *docking protein (SRP receptor)* at the endoplasmic reticulum. Signal sequence and ribosome bind to ER membrane, and translation resumes, with the release of the SRP. The protein being synthesized is concurrently transported into the cisternal space of the ER.

   b. The enzyme *signal peptidase* removes the signal sequence from a protein as it is translated and threaded into the ER cisternal space.

   c. *Glycosyl transferases* add specific carbohydrate groups to specific serines, threonines, and asparagines on the completed protein within the ER.

16 a. Transit sequences target proteins to organelles such as mitochondria and chloroplasts. These transit sequences allow the appropriate proteins to bind receptors on the outer membrane of the organelle and to be imported by them. Signal sequences target proteins to the endoplasmic reticulum for export to the cell surface or extracellular space. Nuclear localization sequences direct proteins destined for the nucleus back to the nucleus. The nuclear localization sequence is recognized at the nuclear pore in the nuclear envelope where the protein that bears it is translocated through the pore into the nucleus.

   b. The retention of nuclear localization sequences after a protein arrives in the nucleus allows proteins that belong in the nucleus to stay or be returned there even after dissolution of the nuclear envelope that occurs during cell division.

17. a.

Cysteine   tRNA. Cys

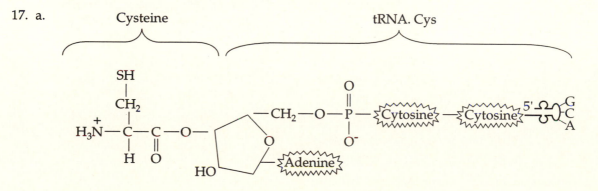

b. The amino acid's amino group is the first to be attached to a growing polypeptide chain. The carboxyl group is bound to the tRNA that brings it to the ribosome. The carboxyl group is involved in a peptide bond only when the next amino acid is brought to the A site.

18. Normally, UGA encodes a translational stop, but in the appropriate nucleotide context, seleno-cysteine, a rare amino acid, is added instead. Several molecules are necessary for this process to occur. The product of the *selC* gene is the tRNA.Ser with the 5' UCA 3' anticodon. This tRNA.Ser is charged with a serine by seryl-tRNA synthetase. The serine on the tRNA is converted to a selenocysteine by the products of the *selA* and *selD* genes. *selB* acts as a special EF-Tu factor to escort the tRNA into the A site of the ribosome when there is a UGA codon within and an appropriate nucleotide context around.

19. a. CCC, CCU, CUC, UCC, CUU, UCU, UUC, UUU

b. proline, serine, leucine, and phenylalanine. This can be determined by consulting the genetic code and determining the amino acids determined by each codon in the copolymer.

c. The percentage of each amino acid can be determined by first calculating the probability of each codon arising by chance given that 75% of the bases are cytosine and 25% of the bases are uracil. The product rule allows us to determine the probability of any given codon being, for example, uracil followed by uracil followed by uracil = $0.25 \times 0.25 \times 0.25 = 0.016$. The probabilities for each codon follow:

CCC Pro $0.75 \times 0.75 \times 0.75 = 0.422$

CCU Pro $0.75 \times 0.75 \times 0.25 = 0.140$

CUC Leu $0.75 \times 0.25 \times 0.75 = 0.140$

UCC Ser $0.25 \times 0.75 \times 0.75 = 0.140$

CUU Leu $0.75 \times 0.25 \times 0.25 = 0.047$

UCU Ser $0.25 \times 0.75 \times 0.25 = 0.047$

UUC Phe $0.25 \times 0.25 \times 0.75 = 0.047$

UUU Phe $0.25 \times 0.25 \times 0.25 = 0.016$

We calculate the probability of an amino acid being encoded by the copolymer by combining the probabilities of codons that encode the same amino acid. For example, CCC and CCU both encode proline, so proline should appear at about $0.422 + 0.140 = 0.562$ or 56.2% of the amino acids encoded by the copolymer.

proline: 56.2%

leucine: 18.7%

serine: 18.7%

phenylalanine: 6.3%

d. An RNA copolymer of 75% uracil and 25% cytosine in random sequence would have codons in the following relative frequencies:

UUU Phe $0.75 \times 0.75 \times 0.75 = 0.422$

UUC Phe $0.75 \times 0.75 \times 0.25 = 0.140$

UCU Ser $0.75 \times 0.25 \times 0.75 = 0.140$

CUU Leu $0.25 \times 0.75 \times 0.75 = 0.140$

UCC Ser $0.75 \times 0.25 \times 0.25 = 0.047$

CUC Leu $0.25 \times 0.75 \times 0.25 = 0.047$

CCU Pro $0.25 \times 0.25 \times 0.75 = 0.047$

CCC Pro $0.25 \times 0.25 \times 0.25 = 0.016$

Calculating as before:

phenylalanine: 56.2%

leucine: 18.7%

serine: 18.7%

proline: 6.3%

e. A copolymer of poly(UC), in which uracil and cytosine alternate precisely, would have two codons, CUC and UCU, in equal frequency. CUC encodes leucine. UCU encodes serine. Both amino acids would occur in equal frequency in the peptides synthesized.

20. a. The genetic code is *continuous* in that there are no intervening bases between codons. That is, codons are as follows: NNN NNN NNN, not NNN N NNN N NNN.

b. The genetic code is *punctuated with start and stop signals* in that there are start codons (AUG in the proper context on a transcript) and stop codons (UGA, UAG, UAA). These signals are intrinsic to the code.

c. The genetic code is *nonoverlapping* in that codons are side by side, sharing no bases with adjacent codons. That is, codons in the sequence ACAGCUCAC could be ACA GCU and CAC, not ACA, CAG, AGC, etc.

d. The genetic code is *degenerate* in that multiple codons can encode for the same amino acid. For example, UCU, UCC, UCA, UCG, AGU, and AGC all encode serine. Knowing that an amino acid in a protein is serine doesn't tell one unequivocally which of the six codons possible actually encoded for that particular serine.

e. The genetic code is *universal* in that almost all organisms encode the same amino acid by a given codon. That is, AUG encodes methionine in bacteria, yeast, pine trees, and people. There are slight differences between organisms separated by hundreds of millions of years of evolution, but otherwise the code is the same.

f. The genetic code is *nonambiguous* in that a given codon will always direct the insertion of the same amino acid in the growing polypeptide. That is, CAU will always encode histidine and not encode, for example, glutamine at other places in the transcript. There are some rare exceptions, one of which is selenocysteine encoded by the UGA codon that normally encodes a "stop."

19

Recombinant DNA Technology and Manipulation of DNA

Problems

1. Restriction endonucleases recognize specific sequences in double-stranded DNA and cut at specific sites within or near the enzyme's recognition sequence.

   a. Distinguish among the following:
      a) DNA with twofold rotational symmetry and palindromic DNA
      b) four cutters and six cutters
      c) type I endonuclease and type II endonuclease
      d) isozymes and isoschizomers
      e) staggered ends and blunt ends
      f) 5′ overhang and 3′ overhang

   b. *E. coli* has a 4.7 Mb genome with a 50% GC content. How many *Eco*RI sites would you expect there to be in the whole *E. coli* genome based on chance occurrence of the *Eco*RI recognition sequence? Approximately how long would the average digestion fragment be if the *Eco*RI enzyme cut at all the *Eco*RI sites?

   c. *Eag*I cuts double-stranded DNA at C^GGCCG (assume sequences to be written 5′ to 3′ unless otherwise labeled), and *Not*I cuts at GC^GGCCGC. Would an *Eag*I-cut fragment and a *Not*I-cut fragment have compatible ends? If not, why not? If so, could an unmethylated *Eag*I-cut fragment ligated to an unmethylated *Not*I-cut fragment be cut by *Not*I? If not, why not. If so, what percentage of the time?

2. The enzymology of restriction enzymes is very precise.

   a. When a restriction endonuclease cuts a molecule of DNA, which of the newly created ends (5′ or 3′) of the DNA retains the phosphate group of the phosphodiester bond?

   b. What enzyme is capable of rejoining two compatible ends of dsDNA? What additional molecules, if any, does the enzyme require? Does the enzyme require that one, both, or neither of the strands being rejoined contain a phosphate group?

3. Plasmids are one of the three major types of vectors typically used for cloning DNA in bacteria such as *E. coli*.

   a. What are the three features required by *E. coli* plasmids that are used for cloning?

   b. Draw a rough map of pUC19 showing the three essential features plus a marker gene for use for insertional inactivation screening.

   c. List three *extra* features that a bacterial plasmid can be engineered with to increase its utility for molecular biologists.

   d. What is the approximate size limit for DNA inserts carried by plasmids such as pUC19?

4. The small weed *Arabidopsis thaliana* has a genome of approximately 100 megabase pairs (Mbps) with approximately 41% GC content.

   a. About how many *Eco*RI restriction sites would you expect to be contained within an average *Arabidopsis* genomic DNA insert (partially) digested with *Eco*RI and inserted in a typical lambda phage vector?

   b. Approximately how many phage would be necessary to carry the entire *Arabidopsis* genome if exactly one genome equivalent of DNA were divided among all the phage?

   c. Approximately how many cosmids would be necessary to carry the entire *Arabidopsis* genome if exactly one genome equivalent of DNA were divided among all the cosmids?

5. In the generation of a genomic library, the method by which genomic DNA inserts are prepared is very important.

   a. Describe three ways of preparing genomic inserts of the appropriate size for the generation of a lambda phage–based genomic library.

   b. What are advantages and disadvantages of each insert preparation method?

6. How many different recombinant cosmids must be present in a *Drosophila* genomic library (diploid genome size: 160 Mb) to have a 99% chance that any given DNA sequence will be represented at least once? Note: you may assume that 40 kb is the optimal insert site.

7. Complementary DNA (cDNA) libraries provide researchers with a source of an organism's or specific tissue's expressed DNA sequences.

   a. Outline the steps of synthesizing double-stranded cDNA. Name the enzymes and reagents used at each step.

   b. Outline the steps of inserting double-stranded cDNA into a lambda phage vector.

   c. What are the advantages of cDNA libraries over genomic libraries?

   d. What are the limitations of cDNA libraries?

   e. What differences would you expect between libraries containing inserts made by oligo(dT)-primed cDNA synthesis and those made by random hexamer-primed cDNA synthesis?

8. Many ways of screening cDNA libraries for genes of interest are known. Briefly describe each of the following methods:

   a. hybridization with a probe corresponding to the gene of interest

   b. hybridization with an oligonucleotide probe

   c. hybridization with a heterologous probe

   d. incubation of an expression library with a labeled antibody

   e. complementation screening

9. Describe the process of labeling DNA using the random primer method. Explain differences in the procedures for labeling the probe radioactively and nonradioactively.

10. How would you use yeast complementation screening in a *ura3 arg1* yeast strain (i.e., mutations at both loci) to identify genes that when mutated cause sensitivity to osmotic shock? Note: neither *ura* nor *arg* is responsible for the sensitivity. The osmotic shock–sensitive strains are killed by exposure to high concentrations of sucrose.

11. Recall that gene expression refers to the amount of RNA made from a particular gene. Northern blot analysis allows one to make important conclusions about the *tissues* in which a gene is expressed, the *level* of expression of the gene in various tissues, the *changes* in the expression of a gene during development, the existence of *splice variants* of a gene, and changes in the expression of a gene after cells are subjected to an experimental procedure such as treatment with a drug.

    a. Describe the process of northern blot analysis.

    b. Answer the following questions with respect to the autoradiogram shown below in which RNA from many sources has been run on an RNA gel, blotted to a membrane, and hybridized to a probe made from an imaginary *Drosophila* gene, *mucin*.

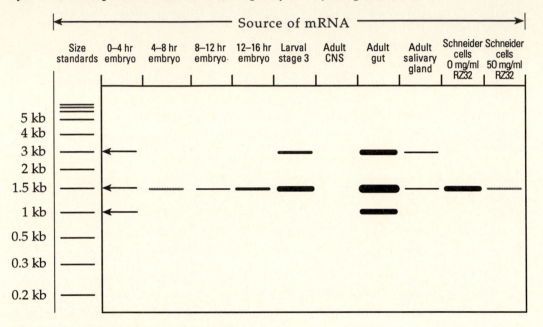

    a. In which tissue(s) is *mucin* expressed?
    b. In which tissue is the expression of *mucin* the highest?
    c. During which developmental stages is *mucin* expressed?
    d. Is there any indication of tissue-specific splicing of *mucin* mRNA?
    e. What happens to *mucin* expression when cells are treated with compound RZ32?

12. Dideoxy sequencing is currently the most widely used method of DNA sequencing and the main technique for sequencing of genomes.

    a. What are the elements of a dideoxy sequencing reaction?

    b. Draw the core structure of dideoxyadenosine triphosphate, deoxyadenosine triphosphate, and adenosine triphosphate; indicate how and where each of them differs from the other.

13. The development of the polymerase chain reaction (PCR) has revolutionized gene cloning and analysis.

    a. What are the essential molecular reagents required by PCR?

    b. What occurs at each of the steps in a typical PCR cycle?

    c. Why are there many cycles in a typical PCR run?

   d. Of the following, which is absolutely required for PCR?
- a) a thermal cycler machine
- b) great thermostability of the DNA polymerase
- c) primers

   e. List five applications of PCR.

14. If PCR were performed on the portion of the *E. coli CysS* gene shown below with the following primers: 5′ TGGTGTTATAGCATAAC 3′ and 5′ CATTTTCATTGGCGCGT 3′, what size product would be expected? Note: only the 5′ to 3′ strand is shown; the actual PCR template would also include the strand complementary to the one below. The number corresponds to the first base of the line; each line's sequence continues on the line below it.

```
  1  tcaacccagt  tcgggtcata  tatagggtgg  tgttatagca  taaccgcacg  atcggatcat
 61  cacgcaatgt  atgctgattc  gcgcgggaaa  tatgggtatt  atacgcaact  caattaccca
121  cacatgtcta  aacggaatct  tcgatgctaa  aaatcttcaa  tactctgaca  cgccaaaaag
181  aggaatttaa  gcctattcac  gccggggaag  tcggcatgta  cgtgtgtgga  atcaccgttt
241  acgatctctg  tcatatcggt  cacgggcgta  cctttgttgc  ttttgacgtg  gttgcgcgct
301  atctgcgttt  cctcggctat  aaactgaagt  atgtgcgcaa  cattaccgat  atcgacgaca
361  aaatcatcaa  acgcgccaat  gaaaatggcg  aaagctttgt  ggcgatggtg  gatcgcatga
```

15. DNA fingerprinting has become an extremely valuable technique for forensics, conservation, and population studies, among other uses.
   a. Distinguish between RFLP, VNTR, and microsatellite analysis.
   b. Explain why it is easier to establish innocence than guilt using DNA fingerprinting.

16. Bertie is married to Matilda, who was previously married to Wynton, now deceased. Wynton and Matilda conceived one child together and adopted one child. Bertie and Matilda have also conceived one child. All members of Matilda's current family have had DNA fingerprinting done at a *single* VNTR locus. Unfortunately, the sheet that identified each child has been misplaced. Identify which fingerprint in each lane (in lanes 5, 6, and 7) correspond to each child.

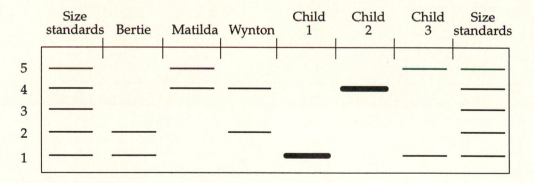

Challenge Probems

17. Given the DNA molecules shown below, how would you construct a pUC19-based plasmid that contains the *trpA* gene in *reverse* orientation with respect to a phage T7 promoter? Available restriction sites are shown in parentheses. The *lacZ* gene in pUC19 has restriction sites inserted in frame within it.

   pUC19: T7 promoter *lacZ* (*Eco*RI *Sma*I *Bam*HI *Sal*I *Bsp*MI *Pst*I *Hind*III)

   lacZ T3 promoter *ori amp*^{R}

   p(trpA): (*Eco*RI *Pst*I) *trpA* (*Sma*I *Hind*III) *ori kan*^{R}

   a. What would you cut each plasmid with to generate the compatible ends needed for directional cloning?

   b. What molecules might you want to *remove* before setting up the ligation reaction?

   c. After transforming *E. coli* with the products of the ligation reaction, what selection criteria would you use to be sure of the following:

      a) The plasmid is pUC19 based and not p(*trpA*) based.

      b) The plasmid contains an insert within the multiple cloning site.

      c) The insert in the multiple cloning site is likely to be *trpA*.

      d) The insert in the multiple cloning site is *trpA* in reverse orientation.

      e) The insert in the multiple cloning site is an unmutated copy of *trpA*.

18. In a restriction mapping experiment using the plasmid P(ST), you cleave the circular plasmid with the individual restriction enzymes *Xba*I, *Asp*718, *Bam*HI, and *Eco*RI individually and in combination with one another and obtain restriction products of the sizes shown below:

ENZYME(S) USED	SIZES OF DNA FRAGMENTS PRODUCED (bp)
*Bam*HI, *Eco*RI, *Xba*I	3,000, 1,550
*Asp*718, *Bam*HI, *Xba*I	3,000, 2,500, 600
*Asp*718, *Bam*HI	5,500, 600
*Asp*718, *Xba*I	3,600, 2,500
*Bam*HI, *Xba*I	3,100, 3,000
*Asp*718	6,100
*Bam*HI	6,100
*Xba*I	6,100

   Using the above data, make a restriction map of the P(ST) plasmid *indicating distances between adjacent restriction sites.*

Team Problems

19. As part of a project to generate a physical map of the mosquito genome, mosquito genomic DNA is cut with *Not*I, a restriction endonuclease that cuts at GC^GGCCGC. A partial digest is done that allows the creation of 100–300 kb genomic fragments.

   a. Draw a YAC clone in which a 200 kb mosquito genomic DNA insert has been inserted.

   b. What auxotrophic strain of yeast could be transformed with the recombinant YACs so that transformants can be distinguished from nontransformants?

   c. How many *Not*I sites on average would be expected with a 100–300 kb insert?

20. The sequence shown below is of the multiple cloning site of an M13 vector. The "universal" M13 sequencing primer is shown above the M13 sequence. The figure below shows an autoradiogram of a dideoxy sequencing gel of a gene fragment cloned into the M13 vector. The universal sequencing primer was used in the sequencing reactions.

   a. Mark on the gel the part of the DNA sequence corresponding to the M13 vector.

   b. What restriction site was the gene fragment cloned into?

   c. What is the sequence of the gene fragment? Label 5' and 3' ends.

M13 universal sequencing primer

5' GTT TTC CCA GAC ACG AC 3'
3' CAA AAG GGT CTG TGC TGC AAC ATT TTG (cont.)

M13 vector near the multiple cloning site

3' CTG CCG GTC ACG GTT CGA ACC CGA CGT (cont.)
3' CCA GCT GAG ATC TCC TAG GGG CCC CGA (cont.)
3' GCT CGA ATT CGT AAT CAT GGT CAT AGC 5'

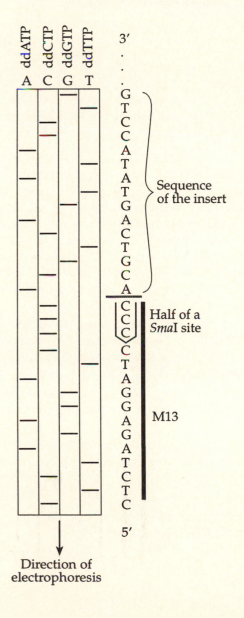

Solutions

1. a.   a) "Twofold rotational symmetry" and "palindrome" are terms often used synonymously by molecular biologists. Twofold rotational symmetry is a characteristic of DNA when the two correponding strands of the duplex have the same base sequence when read with the same polarity (e.g., 5' to 3') Restriction enzymes frequently (but not always) recognize sequences with twofold symmetry. *Bam*HI, for example, recognizes:

   5' GGATCC 3'

   3' CCTAGG 5'

   A palindrome has the same meaning in molecular biology. Linguistically, palindromes are words such as "level" and "radar," which can be read the same forward and backward.

   b) Four-base cutters are restriction enzymes that have a recognition site four-base pairs long. Analogously, a six-base cutter has a six-base-pair recognition site.

   c) Type I endonucleases cut ("^") duplex DNA *outside* of the enzyme's recognition sequence. *Ear*I is an example of a type I endonuclease; its recognition and cut site are:

   5' CTCTTCN^NNN 3'

   3' GAGAAGNNN^N 5'

   Type II endonucleases are more commonly used in molecular biology. Type II endonucleases cut duplex DNA *inside* of the enzyme's recognition sequence. *Eco*RI is an example of a type II endonuclease. Its recognition and cut site are:

   5' G^AATTC 3'

   3' CTTAA^G 5'

   d) Isozymes are *variant forms* of a particular enzyme. Each isozyme catalyzes the same chemical reaction, but varies in the biochemical properties such as pH, salt concentration, or substrate concentration at which they work best. Isozymes are, by definition, encoded by *different* genes within the *same* organism. Isoschizomers are restriction enzymes isolated from *different* organisms that have the same recognition sequence *and* (almost always) the same cut site in duplex DNA. Using isoschizomers is one way to generate compatible ends. Two others are using the same enzyme in both cases and using nonisoschizomers that happen to generate the same "sticky end."

   e) Staggered ends are ends of a duplex DNA molecule that are single stranded for a few bases. When a DNA molecule has blunt ends, there is no single-stranded component; the two strands of the duplex are flush with each other and not staggered or "sticky."

   f) A 5' overhang is a staggered end of DNA in which the strand that protrudes or "overhangs" is the DNA strand that ends (at the overhang) with a 5' carbon exposed. A 3' overhang is a staggered end of DNA in which the strand overhanging terminates with the 3' carbon of the chain exposed.

   b. If *E. coli* has a 4.7 Mb genome with a 50% GC content, the number of *Eco*RI sites that one would expect there to be in the whole genome based on chance occurrence of the *Eco*RI recognition sequence is calculated as follows:

   $p(\text{GAATTC}) = p(\text{G}) \times p(\text{A}) \times p(\text{A}) \times p(\text{T}) \times p(\text{T}) \times p(\text{C})$

   where $p(\text{N})$ = probability of the occurrence of that particular base

   If GC is 50% of the base content, then G = C = A = T = 25% = 0.25.

Thus:

$p(\text{GAATTC}) = (0.25)^6 = 1/4096$ or 1 site every 4,096 base pairs

Number of GAATTC sites $= p(\text{GAATTC}) \times$ number of base pairs total

$$= (1/4096) \times 4,700,000$$

$$= 1147 \text{ } EcoRI \text{ sites}$$

The average digestion fragment length *would* be 4,096 base pairs long.

c. An *Eag*I-cut fragment and a *Not*I-cut fragment *would* have compatible ends because the cut site in both cases yields a 5' GGCC overhang.

An unmethylated *Eag*I-cut fragment ligated to an unmethylated *Not*I-cut fragment could be cut by *Not*I one-quarter of the time, when the *Eag*I-cut fragment happens to have a G 5' to the first C in the recognition sequence.

2. a. When a restriction endonuclease cuts a molecule of DNA, the 5' ends of the cut DNA strands retain the phosphate group of the phosphodiester bond formerly connecting the fragments.

b. Ligase is able to rejoin two compatible ends of dsDNA by forming a covalent bond. In addition to the phosphates on the 5' ends of the DNA strands, ligase requires ATP as the source of energy for phosphodiester bond formation.

3. a. *E. coli* plasmids used for cloning require (1) an origin of replication that is recognized by *E. coli* DNA synthesis initiation proteins, (2) a selectable marker that will allow bacteria that have taken up the plasmid to grow while those untransformed will not, and (3) a restriction site, preferably unique, that allows the linearization of the plasmid and insertion of the piece of DNA to be cloned.

b. A map of pUC19 is shown below. "*ori*" marks the origin of replication. "*amp*ʳ" shows the selectable marker, ampicillin resistance, and "*Eco*RI" indicates the location of the *Eco*RI restriction site within a region called the "polylinker" that contains many unique restriction sites. "*lacZ*" marks the site of the disruptible marker, the gene that encodes for beta-galactosidase.

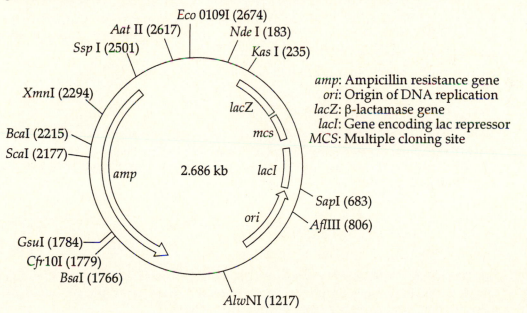

amp: Ampicillin resistance gene
ori: Origin of DNA replication
lacZ: β-lactamase gene
lacI: Gene encoding lac repressor
MCS: Multiple cloning site

c. Some of the extra features that a bacterial plasmid can be engineered with that increase its utility to molecular biologists include the following:

- polylinkers or multiple cloning sites (MCSs) that contain several unique restriction sites clustered in a small region of the plasmid (Polylinkers allow directional cloning and facilitate later cloning.)

- inactivatable markers, such as the *lacZ*+ gene, that allow one to distinguish whether a given bacterial colony has taken up a plasmid *with an insert* (If an insert is successfully ligated at the MCS, the *lacZ* gene is interrupted, converting *lacZ*+ to *lacZ*−, and the bacteria containing the plasmid will be *lacZ*−.)

- origins of replication for eukaryotic cells that allow the plasmid to be replicated in eukaryotes such as yeast

- selectable markers that allow selection for the plasmid in eukaryotic cells

- promoters and enhancers near the cloning site that allow the gene insert to be expressed even if its own regulatory region has not also been cloned into the plasmid

- the origin of replication from filamentous phage that allows the recovery of a *single strand of DNA* from transformed bacteria when they are coinfected with a particular phage

d. The size limit for inserts carried by pUC19 is approximately 5–10 kb. Only fragments of up to a few kilobases are efficiently cloned into plasmid vectors.

4. a. To solve this problem, we need first to calculate the probability of encountering an *Eco*RI site at any position along the length of the *Arabidopsis* genome. *Eco*RI recognizes and cuts at 5′ GAATTC 3′. The probability of encountering a GAATTC sequence by chance is the *product* of the probabilities of encountering each of these six bases. Because the GC ratio is 41% or 0.41, the frequency of G is 0.205. The frequency of C is also 0.205. A and T, therefore, both have probabilities of (1.00 − 0.41)/2 or 0.295. The probability of GAATTC, then, is

$$p(\text{GAATTC}) = p(G) \times p(A) \times p(A) \times p(T) \times p(T) \times p(C)$$
$$= 0.205 \times 0.295 \times 0.295 \times 0.295 \times 0.295 \times 0.205$$
$$= 0.000318 \text{ or } 1/3{,}142 \ (Eco\text{RI sites/bp of } Arabidopsis \text{ DNA})$$

The next part of the question requires that we know the average size of an insert in a typical lambda phage vector. We can use the figures of 10, 15, or 20 kb, which are simply approximations of the capacities of different lambda phages.

So an estimate of the number of *Eco*RI sites in 10, 15, or 20 kb of DNA is:

10 kb: $p(\text{GAATTC}) \times 10{,}000 \text{ bp} = 3.2$

15 kb: $p(\text{GAATTC}) \times 15{,}000 \text{ bp} = 4.8$

20 kb: $p(\text{GAATTC}) \times 20{,}000 \text{ bp} = 6.4$

If we assume that the two ends of the lambda insert are *Eco*RI sites, this accounts for two of the sites in the 10–20 kb interval. That leaves 1–4 sites in a 10–20 kb piece of DNA or 2–3 sites in a 15 kb piece of DNA.

```
•_____•_____•_____•_____•_____•    (• = EcoRI sites if
0          5          10         15         20 kb          spaced equally)
```

b. 100,000,000 bp/10,000 bp per phage = 10,000 phage needed to carry the entire genome if divided equally among all phage. Note: in reality, many more phage would be needed because some sequences are more clonable than others.

c. Given the cosmid insert capacity of 40 kb, only 100,000,000 bp/40,000 bp per cosmid = 2,500 cosmids would be needed.

5. a. Three ways of preparing genomic inserts of the appropriate size for the generation of a lambda phage–based genomic library are (1) sonication of DNA with high-frequency sound, (2) shearing DNA using a syringe needle of an appropriate size, and (3) partial digest with a restriction enzyme.

   b. Sonication and shearing both yield fragments that are broken at essentially random locations. A disadvantage of each is that they both result in blunt and ragged ends. This can be overcome by "filling" in ends with DNA polymerase or snipping off ends using an enzyme specific for single-stranded DNA. "Linkers," or oligonucleotides of a known sequence containing restriction sites to facilitate later cloning, can then be ligated onto the sheared or sonicated fragments.

      An advantage of restriction enzymes is that they yield ends of predictable composition (blunt or staggered). They are not, however, entirely random in their cut sites, so some regions of the genome that, by chance, have very few sites for the enzyme being used will be underrepresented.

6. To have a 99% chance that any given DNA sequence will be represented at least once, N recombinant cosmids must be present in a *Drosophila* genomic library (haploid genome size: 80 Mbp or 80,000 kilobase pair), where N is calculated as follows:

   $$N = ln(1 - P)/ln(1 - f)$$

   $$= ln(1 - 0.99)/ln[1 - (40/80,000)]$$

   $$= -4.605/-0.0005$$

   $$= 9,210 \text{ cosmids}$$

7. a. First isolate mRNA from the tissue and developmental period of choice. Next anneal mRNA with oligo(dT) *primers* (or random hexamers for internally primed cDNA synthesis). Then add *reverse transcriptase, dNTPs,* and *reverse transcriptase buffer* and incubate the reaction at 37°C. The reverse transcriptase polymerizes DNA using an RNA template. This is first strand synthesis. Finally, add *RNaseH, DNA polymerase I, DNA ligase,* and *ATP*. RNaseH degrades the mRNA that is binding to nascent DNA. The DNA polymerase I also removes the ribonucleotides and replaces them with deoxyribonucleotides. The ligase seals the gaps between the last added deoxynucleotide and the deoxynucleotide of the adjoining chain. ATP provides energy for bond formation. This step is called second strand synthesis.

   b. First add linkers (with a restriction enzyme recognition and cutting site within) to the ends of the insert using *T4 ligase*. Then cleave the linkers using the restriction enzyme that cuts within the linkers. Finally, mix the cleaved linkers with the vector that has been cut with the same enzyme. Add ligase to seal the gaps.

   c. DNA in cDNA libraries is (1) stripped of most of the regulatory regions and intronic sequences, (2) representative of the *expressed* genes in a particular tissue at a particular time, and (3) often able to be used as an expression library, in which the processed "minigenes" within are transcribed in the host cell.

   d. cDNA libraries (1) are biased in that they have multiple copies of highly expressed genes and few copies of genes rarely transcribed, (2) provide little regulatory sequence, and (3) are often biased because they contain more 3' ends of genes than full-length cDNA sequences.

   e. An oligo(dT)-primed cDNA synthesis would have a 3' end bias when compared to a random hexamer-primed cDNA synthesis, which can produce a library with a more even distribution of cDNA clones.

8. a. In hybridization with a probe corresponding to the gene of interest, if one already has a fragment of a gene—a genomic fragment, a PCR product, or a partial cDNA (such as the 3′ end only)—one can use these fragments as a template for a labeling reaction. The denatured, labeled product will hybridize to the target cDNA with perfect complementarity over the length of the target that is complementary to the probe.

    b. In hybridization with an oligonucleotide probe, one can synthesize a single-stranded oligonucleotide that base-pairs perfectly with the target DNA over the short length of the oligonucleotide. The strength of hybridization depends on the length and the sequence of the probe.

    c. In hybridization with a heterologous probe, if one probes a cDNA library with a cDNA from another species or another member of the gene family (a paralog) and the stringency of hybridization is not too high, one can identify not only identical, but also related, cDNAs in a library.

    d. In incubation of an expression library with a labeled antibody, expression cloning involves putting cDNAs in plasmids (or phage) with the important feature being that the cDNAs are positioned in frame with a promotor sequence that allows the production of the product of the cDNA within the organism transformed (or transfected) with the expression construct. Colony/plaque transfer is then done as usual, and the filters are incubated with antibodies that will recognize the correct clone by virtue of its producing the protein encoded by the cDNA. Therefore, antibody is concentrated at the site of the correctly transformed plaque or colony.

    e. In complementation screening, mutant organisms are transformed with cDNAs (or candidate cDNAs) in transformation vectors. Those that are rescued (returned to wild type) by the addition of a DNA are likely to contain a copy of the gene that is mutated along with an adequate control region to allow its expression.

9. Random DNA labeling involves (1) single-stranded (denatured) DNA for use as a template, (2) a mixture of random hexanucleotides (ACTTCG, as one example) for use as primers of the DNA polymerization reaction, (3) dNTPs as the building blocks and energy source for probe synthesis, and (4) labeled dNTP ($\alpha[^{32}P]dATP$ or dUTP-biotin or dUTP-digoxigenin) to allow visualization of hybridization events (by exposure of X-ray film or colorimetrically, respectively) and buffer (with magnesium, salt, and buffering agents) to allow the polymerase to function optimally.

10. Transform osmotically sensitive yeast with yeast transformation vectors containing genomic DNA fragments. Plate the transformants on a concentration of sucrose high enough to cause osmotic shock (and death) in the mutants, but not in the wild-type yeast. Those yeast transformed with DNA fragments that rescue their osmotic pressure sensitivity phenotype will grow. Isolate from these rescued yeast the transformation vector and the genomic DNA fragments they contain, which are the genes required for a wild-type osmotic shock phenotype.

11. a. Northern blot analysis involves RNA from any source being separated on an RNA gel (usually containing denaturant), blotted to a membrane by capillary transfer or electrotransfer, and hybridized to a labeled probe. In this way, RNA complementary to the probe can be detected and quantitated, its size determined, and any variations in size detected.

    b. With respect to the autoradiogram shown:
    *   *Mucin* is expressed in the adult gut and salivary glands.
    *   *Mucin* expression is highest in the gut.
    *   *Mucin* is expressed in *Drosophila* from the 8–12 hour embryo stage through adulthood.
    *   *Mucin* mRNA seems to be spliced differently from the larval stage onward. There is also a tissue-specific splice form in the adult gut.

- Compound RZ32 decreases the total amount of *mucin* mRNA in Schneider cells, a type of *Drosophila* cell that grows in tissue culture.

- To analyze tissue-specific effects of RZ32 requires one to isolate mRNA from each of the specific tissues of interest from RZ32-treated flies and to resolve samples from each tissue on a gel along with mRNA from untreated tissue.

12. a. The basic elements of the manual dideoxy sequencing reactions are (1) single-stranded DNA to serve as a template for the synthesis of DNA sequencing fragments, (2) DNA polymerase without 3' to 5' exonuclease activity, (3) a single primer to prime the DNA polymerase at a specific site, (4) dNTPs to provide monomers for DNA synthesis and energy for the polymerization reaction, and (5) ddATP or ddCTP or ddGTP or ddTTP (a different one of these in each of four separate sequencing reactions) to provide the base-specific chain terminator that allows the generation of fragments of *informative* lengths in each of the four reactions.

b.

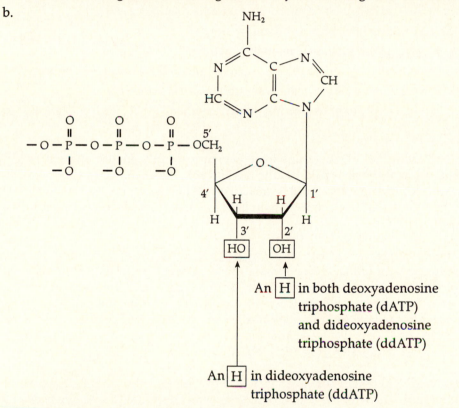

An $\boxed{\text{H}}$ in both deoxyadenosine triphosphate (dATP) and dideoxyadenosine triphosphate (ddATP)

An $\boxed{\text{H}}$ in dideoxyadenosine triphosphate (ddATP)

13. a. The essential molecular reagents required by PCR are (1) template DNA (single or double stranded), (2) two primers, both in excess quantity, with their target sequence within the template DNA and oriented with 3' ends pointing toward each other, (3) all four dNTPs to provide the monomers and energy for DNA synthesis, (4) a DNA polymerase, preferably thermostable, and (5) a buffer containing magnesium to provide optimal reaction conditions for the DNA polymerase.

b. A typical PCR cycle has three segments: (1) denaturation, when the strands of double-stranded DNA separate, (2) annealing, when primers bind to target sequences in preparation for directed DNA synthesis, and (3) extension, when DNA polymerase extends the primers using the dNTPs included in the reaction.

c. PCR is a chain reaction. To increase from one target molecule to over a million (2^{20}) would require 20 cycles (or doublings) if the reaction worked with perfect efficiency. The reaction doesn't work with perfect efficiency, however. Scientists have found that 25 to 30 cycles typically provide the maximal yield for a given reaction volume. The limiting factor is generally either the dNTP concentration or the primers (both of which diminish throughout the reaction) or the *Taq* polymerase (which gradually loses its activity throughout the reaction).

d. Of the following, only the primers are absolutely required for PCR.

- A thermal cycler machine is dispensable. Early on, people manually transfered tubes among different water baths.

- Great thermostability of the DNA polymerase is also dispensable. One could add new DNA polymerase after each reaction, though this, too, would be inefficient.

- Primers are required for all polymerase chain reactions because DNA polymerase is dependent on a 3' hydroxyl group for the initiation of DNA synthesis.

e. 1. DNA fingerprinting in forensics and anthropology

2. detection of VNTR polymorphism in gene mapping

3. identification of the DNA flanking a particular known DNA sequence (inverse PCR)

4. genetic testing to determine if a genome contains a particular allele

5. generation of large quantities of any short length of DNA for applications such as cloning

14. The PCR products would be 360 base pairs long. One measures the length of the product from where the 5' end of each primer would bind to the template (28–388). All of the bases between would be synthesized. The primers would be included within one or the other of the two strands at the ends of the product. Primers/primer templates are underlined below. The amplified portion is in italics.

```
  1 tcaacccagt tcgggtcata tatagggtgg tgttatagca taaccgcacg atcggatcat
 61 cacgcaatgt atgctgattc gcgcgggaaa tatgggtatt atacgcaact caattaccca
121 cacatgtcta aacggaatct tcgatgctaa aaatcttcaa tactctgaca cgccaaaaag
181 aggaatttaa gcctattcac gccggggaag tcggcatgta cgtgtgtgga atcaccgttt
241 acgatctctg tcatatcggt cacgggcgta cctttgttgc ttttgacgtg gttgcgcgct
301 atctgcgttt cctcggctat aaactgaagt atgtgcgcaa cattaccgat atcgacgaca
361 aaatcatcaa acgcgccaat gaaaatggcg aaagctttgt ggcgatggtg gatcgcatga
```

15. a. In restriction fragment length polymorphism (RFLP) analysis, one examines the natural variations among individuals in the length of fragments of DNA after they have been cut with restriction endonucleases. The various DNA are detected using nucleic acid probes specific to them. These polymorphisms (RFLPs) are generally a consequence of DNA rearrangements or other mutations that create, delete, or move restriction sites from their usual locations. Sometimes they are due to nearby repetitive DNA that is variable from individual to individual in the number of repeats that are present. RFLPs have been used to map genes and to make genetic fingerprints.

VNTR analysis is a technique that takes advantage of the fact that repetitive DNA is common in many eukaryotic genomes and the *number* of tandem, or adjacent, repeats (e.g., of the bases "CCGACGGCAGA") can vary widely from individual to individual and even between chromosomes within an individual. In VNTR analysis, each of the tandemly repeated oligonucleotide sequences detected is approximately 11–60 bp long. Oligonucleotide probes to the particular VNTR sequence in question are hybridized to Southern blots of restriction enzyme–cut genomic DNA to reveal the size of each "allele" at the particular VNTR loci being examined. VNTRs are also known as minisatellites.

When the repeat unit length is small (2, 3, or 4 nucleotides) and the number of repeats is highly variable from chromosome to chromosome (that is, the locus is polymorphic), then the locus is called a microsatellite. Analysis of microsatellites, which tend to be more evenly spread throughout the genome than minisatellites, can be done by PCR using primers designed to hybridize to the nonvariable regions flanking the microsatellite. Using PCR allows one to type very small quantities of DNA.

b. The presence of alleles (markers) that are *not* present in the suspect definitively excludes the suspect as the source of the DNA at the crime scene (or DNA that has been linked to the crime). If the markers on the DNA found at the scene of the crime match those of the suspect, then it still remains a question whether it is simply a coincidence (both the suspect and the true perpetrator happen to share the same marker alleles at the loci examined) or whether it is a consequence of the fact that the suspect is the perpetrator. By testing several marker loci, one can reduce to *nearly* zero the probability that any observed similarity between suspect and DNA from the crime scene is due simply to chance.

16. First we identify the *genotype* of each individual from the Southern blot:

    Bertie: 1, 2　　Child 1: 1, 1

    Matilda: 4, 5　　Child 2: 4, 4

    Wynton: 2, 4　　Child 3: 1, 5

The child that Bertie and Matilda conceived must have one allele from each parent. Child 1 does not have an allele from Matilda. Child 2 does not have an allele from Bertie. Only child 3 possesses one allele from Bertie and one from Matilda: 1, 5. Child 3 is therefore the best candidate for the child conceived by Bernie and Matilda.

Child 2 is the only child who has one allele from both Wynton and Matilda: 4, 4. Therefore, child 2 is the best candidate for the child conceived by Wynton and Matilda.

Child 1, genotype 1, 1, is the only child who does not share an allele with Matilda. Though this *could* be a child of Bertie's, this is the best candidate for the child who was adopted by Wynton and Matilda.

17. a. Cut each plasmid with *Pst*I and *Sma*I to generate the compatible ends needed for directional cloning. These restriction enzymes will cut the p(*trpA*) in such a way as to put the downstream part of *trpA*⁺ gene next to the T7 promoter. This *reverse* insertion will allow the RNA polymerase binding to the T7 promoter to transcribe from the nontemplate (sense) strand, making an *antisense* RNA molecule.

    b. If the "stuffer" fragment of pUC19 and the remainder of the p(*trpA*) vector were removed before ligation, one wouldn't have to worry about competing ligations between those fragments, which also have compatible cut ends. Removing the restriction enzyme is also *very* important; it would otherwise digest the newly ligated recombinant plasmid.

    c. After transforming *E. coli* with the products of the ligation reaction, several selection criteria can be used to distinguish the correct clone from the others.

       • Add ampicillin to the agarose plates onto which the transformation mixture is plated. Only the cells containing plasmids conferring resistance to ampicillin will grow. pUC19 carries *amp*ʳ. p(*trpA*) carries only *kan*ʳ, so the cells containing only this plasmid will not survive.

       • Perform blue-white selection based on the activity of beta-galactosidase. Plasmids without inserts have a functional (uninterrupted) *lacZ*⁺ gene. Bacteria with these plasmids turn blue when plated on media with X-gal, a galactose derivative. Plasmids *with* inserts that disrupt the *lacZ*⁺ gene will not make beta-galactosidase, so the colonies will remain white.

- Make purified preparations of the plasmid from the "white" bacteria that grow on the ampicillin-containing agarose. Cutting the plasmid with the enzymes used in the original cloning (*Sma*I and *Pst*I, in this case) should release an insert *of the appropriate size*. The size of the insert can be determined by running the digestion product on an agarose gel along with uncut DNA and size standards.

- Cut the plasmids with one of the restriction enzymes used in the initial cloning and one other enzyme that *cuts asymmetrically (near one end or the other) within the insert, but not within the vector*. The digestion will yield either a small or a large fragment of the insert. The size of the fragment can be assessed on an agarose gel and correlated to the size expected from the enzymes used.

- This requires DNA sequencing of the insert. Using primers designed to hybridize to the edge of the polylinker and other primers designed to hybridize to sequences within the insert, one can carry out sequencing reactions that will assess the complete DNA sequence of both coding and template strands.

18. One way to make a restriction map of a plasmid is to cut samples of the plasmid with two or more restriction enzymes in various combinations as is described in this problem. The generation of a restriction map from the data can be somewhat involved, so it is best to proceed in a step-by-step manner, checking off restriction data as you proceed. When data become very involved, restriction maps are best made with the assistance of a computer. However it is done, mapping involves the following steps:

1. Verify that all DNA is accounted for by adding up the length of all restriction fragments. In this case, the sizes of the bands in all but one digestion reaction add up to 6,100 bp.

2. For any that don't add up to what the others do, find out how much is missing, and account for it. In the *Bam*HI, *Eco*RI, and *Xba*I triple digest, we are missing 1,550 bp. It is easy to imagine that each plasmid in this reaction must be broken into a 3,000 bp piece and two 1,550 bp pieces, which, because they run together on a gel, would appear as one thicker band.

3. Draw a map without sites on it first. If a linear piece of DNA is being mapped, draw a line and mark one end "0" and one end whatever the full length of the piece of DNA is. If a circular piece of DNA is being mapped—as is the case here because we are mapping a plasmid—draw a circle, and write the name of the plasmid and its size in bp or kb in the center of the circle.

4. Put one restriction site, say *Xba*I, anywhere on the circle's edge, and check off the line of restriction data that corresponds to the *Xba*I digestion.

5. Find a line of restriction data in which *Xba*I appears, for example, the *Xba*I/*Bam*HI digest that yielded a 3,100 bp and a 3,000 bp fragment. It is generally best to choose the line with the *fewest* bands generated.

6. Put the second restriction site, *Bam*HI in this case, on the circle at the approximate distance from the first site that the double digest products dictate. In this case, 3,000 bp one way is very close to 3,100 bp the other way, so we can put the *Bam*HI site almost directly across the circle from *Xba*I.

7. Put "3,000 bp" and "3,100 bp" on the circle directly between the two sites, and check if the lines of restriction data corresponding to the *Bam*HI/*Xba*I double digest *and* the line corresponding to the *Bam*HI single digests.

8. Now comes the trickier part. Go to another double digest involving one of the two enzymes you've mapped so far and another enzyme. In this case, we can pick either *Asp*718/*Xba*I or *Asp*718/*Bam*HI. Let's pick the first one. We know where *Bam*HI is. We now know that *Asp*718 is 600 bp away. But which way? To the left or right? Mark in pencil the two possibilities, noting to yourself that only one of the two locations really has the *Asp*718 restriction site.

9. Next, look at the triple digest: *Asp*718/*Xba*I/*Bam*HI. The fragments that are a product of that digest should resolve the question of where the *Asp* site is. We can see in that line of data, for instance, that the 3,100 bp band of the *Bam*HI/*Xba*I digestion is absent. That suggests it was cut by *Asp*718. So we go with the idea that *Asp*718 is 600 bp from *Bam*HI and on the 3,100 bp side.

10. Erase the other alternative location, write the 600 bp and 2,500 bp sizes on the appropriate location on the map, and check off the *Asp*718 single digest and the triple digest.

11. Verify the map is correct so far by checking the sizes of the bands generated by the *Asp*718/*Bam*HI double digest. The data are consistent with our map, so we can check off that line of data.

12. The only line of data remaining unincorporated into the map is the first line, the *Bam*HI/*Eco*RI/*Xba*I triple digest. The *Eco*RI site is not yet on the map. It is easy to see now, though, that *Eco*RI simply cuts the 3,100 bp fragment in half to give two 1,550 bp fragments. Mark the *Eco*RI site and midway between the two ends of the 3,100 bp *Bam*HI/*Xba*I arc of the plasmid. Write in the 1,550 bp sizes. Check off the triple digestion that involves *Bam*HI, *Eco*RI, and *Xba*I.

13. Check the map, clean it up, and you are through!

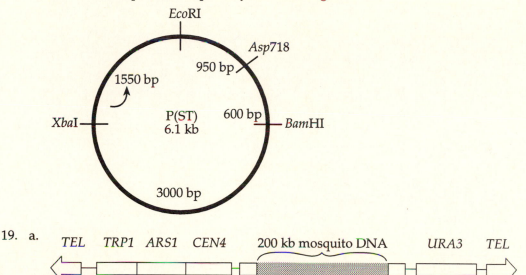

19. a.

TEL   TRP1   ARS1   CEN4   200 kb mosquito DNA   URA3   TEL

b. A *trp⁻ ura⁻* strain of yeast would allow those yeast that take up a recombinant YAC to be distinguished from those that have not. The YAC is composed of one arm that is *ura⁺* and one arm that is *trp⁺*. Both arms are needed. TRP1 and URA3 are the selectable markers for many yeast artificial chromosomes.

c. Assuming the relative frequency of each base is 0.25, the probability of GCGGCCGC occurring is $(0.25)^8$ or 1/65,536. Therefore, there should be between 1.5 (100,000 × 1/65,536) and 4.6 (300,000 × 1/65,536) or, on average, three *Not*I sites per insert, including the ones at the ends.

20. a. The part of gel showing the DNA sequence corresponding to the M13 vector is labeled "M13" and marked by a bar.

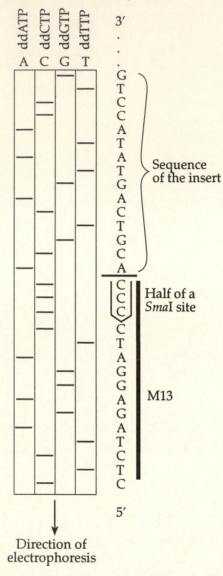

Direction of
electrophoresis

b. The gene fragment was cloned into GGG/CCC (*Sma*I site).

c. The sequence of the portion of the gene fragment that we can determine from the gel shown is 5' ACG TCA GTA TAC CTG ... 3'.

Some of the M13 sequence is not shown here because it is not part of the cloned gene.

```
5' GTT TTC CCA GAC ACG ACG TTG TAA AAC GAC GGC CAG TGC CAA GCT TGG GCT GCA
3' CAA AAG GGT CTG TGC TGC AAC ATT TTG CTG CCG GTC ACG GTT CGA ACC CGA CGT

                                5' ACG TCA GTA TAC CTG ... (more insert sequence ... )
5' GGT CGA CTC TAG AGG ATC CCC
3' CCA GCT GAG ATC TCC TAG GGG CCC CGA GCT CGA ATT CGT AAT CAT GGT CAT AGC
```

Note: M13 sequence that will be found on the *other* side of the inserted DNA is indicated by dashes (---) underneath the bases.

20
Prokaryotic Gene Regulation

Problems

1. a. Compare the characteristics of constitutive genes (or housekeeping genes) and regulated genes.
   b. How could one determine experimentally where a given gene is located on the spectrum between constitutive and regulated genes?

2. What distinguishes effectors, inducers, and transcriptional regulatory proteins from each other? Provide an example of each.

3. Why are genes in an operon for the most part coordinately expressed?

4. Draw the organization of the *lac* operon, labeling the genes and gene products where applicable.

5. Assuming all other genes are wild type, what effects would you expect to arise as a result of each of the following mutations in a normal *E. coli* cell?
   a. *Plac*$^-$
   b. *lacO*^{c}
   c. *lacZ*$^-$ (missense mutation)
   d. *lacZ*$^-$ (polar/nonsense mutation)
   e. *lacY*$^-$ (missense mutation)
   f. *lacA*$^-$

6. Illustrate how catabolite repression (the glucose effect) modulates *lac* operon gene expression. Label all molecules involved and the consequences of the molecular interactions that occur.

7. When an *E. coli* is starving, intracellular cAMP concentrations are high, many alternative sugar operons are derepressed, and necessary enzymes are produced to make use of the alternative sugars. When glucose is available and being taken up by the cell, however, cAMP levels fall, and the expression of genes encoding enzymes that metabolize alternative sugars is significantly reduced. Describe the model of how low glucose levels within an *E. coli* cell effect the derepression of operons involved in the catabolism of other sugars.

8. On the sequence of the *lac* operon regulatory region and upstream transcribed sequence shown below, identify the following features:
   a. start of transcription (mark as → start of transcription)
   b. −10 and +1 bases
   c. site protected by the product of the *lacI* gene (mark as protected by *lac* repressor)
   d. site(s) bound by the CAP-cAMP complex (mark as CAP binding site)
   e. site(s) of RNA polymerase complexed with sigma factor (mark as −35 and −10)
   f. Write the bases of the transcript below the double-stranded DNA (mark 5′ and 3′ end).

g. Write in the bases of the 16S rRNA that bind to the Shine-Dalgarno sequence.

h. the part of the leader sequence of *lacZ* mRNA that directs ribosomes to the start of translation.

i. *lacZ* start of translation (start of translation)

```
5' ..GCGCAACGCAATTAATGTGAGTTAGCTCACTCATTAGGCACCCCAGGCTTTACACTTTATGCTTC
3' ..CGCGTTGCGTTAATTACACTCAATCGAGTGAGTAATCCGTGGGGTCCGAAATGTGAAATACGAAG
                                                    (continued below...)
5' CGGCTCGTATGTTGTGTGGAATTGTGAGCGGATAACAATTTCACACAGGAAACAGCTATGACCAA..
3' GCCGAGCATACAACACACCTTAACACTCGCCTATTGTTAAAGTGTGTCCTTTGTCGATACTGGTT..
```

9. How is a repressible operon different from an inducible operon? Give examples of each.

10. Describe the precise molecular conditions that give rise to each of the three alternative RNA stem loop structures that can form during transcription/translation of the *trp* leader region.

11. What effects would you expect from the following mutations in the *trpL* gene? In the numbering below, +1 corresponds to the first base of the *trpL* gene to be transcribed.
    a. 56 (G → U) and 59 (G → U)
    b. 117 (C → U) or 132 (G → A)
    c. deletion of 6 bases (54–59) and 69 (U → G)
    d. deletion of 6 bases (54–59) and 71 (A → G)

12. Briefly describe how tryptophan metabolism is regulated at the transcriptional, transcriptional-translational, and protein levels.

13. What is the difference in *outcome* for an *E. coli* cell if lambda phage enters the lytic pathway rather than the lysogenic reproductive pathway?

14. The eukaryotic genome is very fragmentary in its organization, and the lambda phage genome is much more organized and compact. On the sketch of the circularized genome of lambda phage shown on the following page, mark the following:
    a. early transcribed genes
    b. genes involved in the lytic pathway
    c. genes involved in the lysogenic pathway
    d. genes involved in the production and assembly of the phage coat (head and tail)

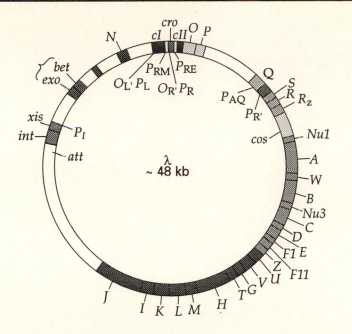

15. What early situation or event flips lambda's genetic switch toward lysogeny? Toward lysis? What contribution does the cell make to the decision?

16. If a lambda phage infects *E. coli* that was lysogenic for lambda, would you expect the incoming lambda phage to be able to lyse the cell already lysogenic for the earlier entering lambda phage?

Challenge Problems

17. List the lambda genes in their *order of expression* upon lambda's entry into a cell when:
    a. the decision is made to follow the *lysogenic* path.
    b. the decision is made to follow the *lytic* path.

18. How does the O_R mediate each of the following (if it does):
    a. promotion of both rightward gene expression and leftward gene expression
    b. promotion of rightward gene expression and repression of leftward gene expression
    c. promotion of leftward gene expression and repression of rightward gene expression
    d. repression of both rightward and leftward gene expression

Team Problems

19. What would be the effect of the following mutations on the quantity of tryptophan synthetase activity and the accumulation of intermediates in the *trp* pathway in *E. coli*?
    a. *trpA⁻*
    b. *trpB⁻* (missense mutation)
    c. *trpC⁻* (polar mutation)
    d. *trpR⁻*
    e. *trpRˢ* (superrepressor mutation)
    f. *trpO⁻*

20. Bacteria can be made to be transiently and partially diploid (merozygous) by the process of con-jugation. While the bacteria are in the diploid state, one can examine the effects of mutations in *cis* (on the same molecule of DNA) and in *trans* (on different molecules of DNA). What would be the effect on the expression of *lacZ, lacY,* and *lacA* genes in cells of the following genotypes in the *presence* and *absence* of lactose? You may assume glucose levels are low.

a. $lacI^+ \ lacO^+ \ lacZ^- \ lacY^+$
   $lacI^- \ lacO^+ \ lacZ^+ \ lacY^-$

b. $lacI^+ \ lacO^+ \ lacZ^- \ lacY^+$
   $lacI^s \ lacO^+ \ lacZ^+ \ lacY^-$

c. $lacI^+ \ lacO^+ \ lacZ^- \ lacY^+$
   $lacI^{-d} \ lacO^+ \ lacZ^+ \ lacY^-$

d. $lacI^s \ lacO^+ \ lacZ^- \ lacY^+$
   $lacI^{-d} \ lacO^+ \ lacZ^+ \ lacY^-$

Solutions

1. a. In prokaryotes, constitutive genes (1) are always transcriptionally active, though not always highly active, and (2) tend to encode proteins with cell maintenance functions—protein syn-thesis and sugar metabolism, for example. Regulated genes (1) are much more variable in their level of transcription, with it sometimes very low and sometimes very high, and (2) tend to encode proteins with occasionally needed functions such as amino acid biosynthesis or biosynthesis of a rare polysaccharide.

   b. To determine whether a gene is constitutively expressed, you need to show that it is tran-scribed in all tissues throughout the organism's life. A probe can be made from the gene and hybridized either to tissue from the organism (*in situ* hybridization) or to RNA prepared from many tissue types and many developmental stages that has been separated by gel elec-trophoresis and transferred to a nylon membrane (northern analysis).

2. Effectors are *small* molecules (cAMP and allolactose, for example) that control the expression of many regulated genes by binding to transcriptional regulatory proteins. They either induce expression by binding (such as allolactose that binds to *lac* repressor protein) or inhibit expression.

   Inducers are the subset of effector molecules that promotes transcription by binding transcrip-tion regulatory factors. Allolactose is an example of an inducer.

   Transcriptional regulatory proteins are a broad group of proteins that activate or repress tran-scription from a promoter. *Lac* repressor protein and catabolite response protein (CRP) are both transcriptional regulatory proteins.

3. An operon is a cluster of related genes whose expression is regulated together (coordinately) by a promoter, operator, and operator-regulator protein interactions. Because the genes in the operon share the same transcriptional regulation machinery, they are under the same tran-scriptional control.

4. $\underline{P_{lac} \ lacO \ lacZ \qquad\qquad lacY \qquad\qquad lacA}$
   $\quad$ beta-galactosidase $\quad$ permease $\quad$ transacetylase

5. a. With a defective promoter, no beta-galactosidase, permease, or transacetylase will be produced under any circumstances.

   b. Without a functional operator, the three genes encoded by the *lac* operon will *not* be under inducible control by allolactose and the *lac* repressor. Though the genes *might* not be transcribed as actively (depending on the nature of the mutation), they will be transcribed nearly constitutively—whenever the level of glucose is low and the catabolite gene activator protein (CAP) is bound to cAMP.

   c. With a missense mutation in the *lacZ* gene, one would expect production of a defective beta-galactosidase protein, but normally regulated production of permease and transacetylase.

   d. With a polar nonsense mutation in the *lacZ* gene, no beta-galactosidase, permease, or transacetylase will be produced under any circumstances. The ribosomes leave the *lacZ* gene prematurely at the stop codon and don't slide to the downstream genes in operon.

   e. With a missense mutation in *lacY*, no permease will be made, but beta-galactosidase and transacetylase will be produced.

   f. No matter which kind of mutation occurs in *lacA*, the result will simply be production of no transacetylase or defective transacetylase.

6. *crp* gene → CAP monomer → CAP = catabolite gene activator protein

   CAP + cAMP = active positive-regulation complex

   CAP-cAMP complex binds to the CAP site in the *lac* promoter (P_{lac}), bending the DNA, which facilitates interaction between the CAP transcription activation domain and RNA polymerase. If the repressor has been induced to leave the operator by the induced allolactose via the *lac* operator site, the RNA polymerase can then transcribe the *lac* operon genes.

7. When glucose is being transported into the cell, many enzymatic changes occur, including the dephosphorylation of an enzyme known as III^{Glc}. Phosphorylated III^{Glc} *deactivates* adenylate cyclase. Adenylate cyclase normally catalyzes the conversion of ATP to cyclic AMP (cAMP). So *deactivation* of adenylate cyclase reduces the amount of cAMP in the cell. Phosphodiesterase further contributes to the depletion of cAMP because it degrades cAMP into 5' AMP. The cAMP that normally would bind to CAP protein to give rise to the transcription activator CAP-cAMP is largely unavailable. The *lac* operon, and *many* other sugar catabolism operons as well, depend on CAP-cAMP for maximal transcription. In this way, the relationship in which high cAMP levels (during glucose starvation) give rise to high expression of alternative sugar catabolism operons is generated. Low cAMP levels (during glucose abundance) lead to reduced expression of alternative (and energetically less efficient) sugar catabolism pathway genes.

8.
        (d) CAP binding site   **(e) –35**

5'..GCGCAACGCAATTAA<u>TGTGA</u>GTTAGC<u>TCACT</u>CATTAGGCACCCCAGGCT<u>TTACA</u>CTTTATGCTTC
3'..CGCGTTGCGTTAATTACACTCAATCGAGTGAGTAATCCGTGGGGTCCGAAATGTGAAATACGAAG
        Protected by CAP   Protected by RNA (cont.)

    (b, e)   **(a, b) start of transcription ...**
    –10   **–4** **+1 →**   **+21**

5' CGGCTCG<u>TATGTT</u>GTGTGGAATTGTGAGCGGATAACAATTTCACAC<u>AGGA</u>AACAGCT<u>ATG</u>ACCAA..
3' GCCGAGCATACAACA<u>CACCTTAACACTCGCCTATTGTTAAA</u>GTGTGTCCTTTGTCGATACTGGTT..
  polymerase/σ   **(c) protected by lac repressor**

(g) 16S rRNA   (i) Start of translation

(f)   3' ..AU<u>UCCUCC</u>AU..

mRNA (transcript)   5' AAUUGUGAGCGGAUAACAAUUUCACAC<u>AGGA</u>AACAGCU<u>AUG</u>ACCAA...

                                          (h) Shine-Dalgarno

protein (β-galactosidase)   NH$_2$ fMetThrAsn...

9. Repressible operons are those whose expression levels are reduced by the addition of certain chemicals to the medium. An example is the tryptophan operon, which is repressed when tryptophan is provided to the cell. Inducible operons are those whose expression levels are increased by the addition of a certain chemical to the medium. The lactose operon is an example of an inducible operon in that the addition of lactose induces higher expression. (There is also an indirect repressible component to the *lac* operon: when glucose is added to the medium, the expression of *lac* operon genes is repressed by the loss of the CAP-cAMP complex.)

10. The pause structure results from the pairing of region 1 and region 2 of the leader region. These regions pair because of their complementary sequence and serve to delay further *transcription* and thereby allow ribosomes to load onto the nascent transcript and travel on the mRNA right next to the RNA polymerase that is making it. The pause structure, then, gives rise to *tightly coupled transcription and translation*.

The antitermination structure results from the pairing of region 2 and region 3 of the leader region. These regions pair because of their complementary sequence and because of two tryptophan codons in region 1, just preceding region 2; pairing causes the ribosome to delay threading that stretch of mRNA through itself while waiting for a tryptophan-charged tRNA (Trp-tRNA.Trp). If there is little tryptophan in the cell, there will also be little Trp-tRNA.Trp, making the pause very long. During that pause, regions 2 and 3 pair to make the antitermination structure, and the RNA polymerase can move ahead to transcribe the *trpE* open reading frame. If there is enough tryptophan in the cell, there will also be enough charged tryptophan tRNA. With enough tryptophan tRNA, there will be no pause, so allowing regions 2 and 3 will not form a stable antitermination stem loop.

The termination stucture results from the pairing of region 3 and region 4 of the leader region. These regions pair because the ribosome pauses at the stop codon at the end of the *trpL* gene. If the ribosome did not pause at the two tryptophan codons (because tryptophan-charged tRNAs were abundant), then the 3–4 termination structure (the attenuator) is right behind the RNA polymerase. This attenuator structure has the ability to stop transcription in a manner similar to *rho*-independent transcription termination.

11. a. unattenuated *trp* operon transcription when *cysteine* (not tryptophan) levels are low

b. reduced efficiency of transcription termination at the attenuator because the region 3–4 termination structure is destabilized

c. reduced formation of antitermination structure because the two tryptophan codons are deleted and reduced formation of termination structure because the stop codon (69–71) has been converted to a codon for glycine (Also two tryptophans would be deleted from the *trpL* product, and nine amino acids would be added to the C-terminus of the *trpL* product because the next available stop codon is at 96–98. There would be no attenuation.)

d. reduced formation of antitermination structure because the two tryptophan codons are deleted and the formation of the termination structure when tryptophan levels are low (!) instead of high because the stop codon that causes the pause that gives rise to the termination structure is now a tryptophan codon (If tryptophan levels were high, transcription would continue onto the *trpE* gene. Translation of the *trpL* product would continue through the tryptophan codon to add a total of nine amino acids to the C-terminus of the peptide, even though two tryptophan amino acids would be missing from the peptide due to the deletion of the two tryptophan codons.)

12. Regulation of tryptophan metabolism at the transcriptional level occurs by the binding of tryptophan to the *trpR*-encoded *trp* repressor to reduce transcription of the *trp* operon when tryptophan levels are high.

    Transcriptional-translational regulation of tryptophan metabolism occurs by the attenuation of transcription of genes in the *trp* operon when tryptophan levels are high (as measured by Trp-tRNA.Trp availability).

    Protein-level regulation of tryptophan metabolism occurs by end-product inhibition of the first enzyme in the tryptophan biosynthesis pathway (anthranilate synthetase) by tryptophan. When tryptophan is abundant, it binds to the enzyme, causing a conformational change—an allosteric shift—that prevents the enzyme from catalyzing the first step in the synthesis pathway.

13. The lysogenic pathway results in the lambda phage DNA integrating into the *E. coli* genome and lying dormant there, replicating with the *E. coli* chromosome, and excising when cellular conditions change for the worse. The lytic pathway results in the lambda phage DNA being replicated independent of the *E. coli* chromosome, phage coat protein being made, and the phage being assembled and released from the cell, destroying (lysing, or bursting open) the cell in the process.

14.

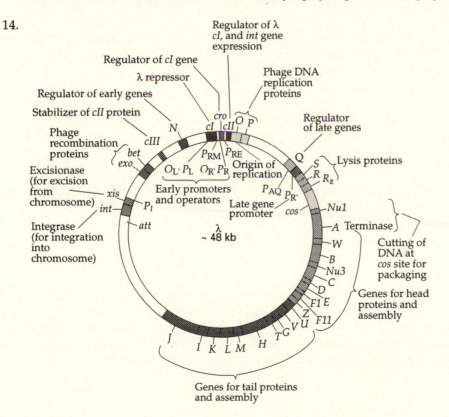

15. The early event that switches lambda phage's genetic switch to lysogeny is the rapid increase in abundance of the protein products of *cII* (right early operon) and *cIII* (left early operon) genes. The *cII* protein is stabilized by the *cIII* protein and activates transcription of the *cI* gene from P_{RE} in the clockwise direction. The product of *cI* is the lambda repressor protein that blocks the expression of genes necessary for DNA replication, phage assembly, and cell lysis. If conditions are good for the infected bacteria (it is metabolizing, transcribing, and dividing rapidly), then the repressor protein will quickly be in abundance, and lysogeny will be favored.

    If conditions are not good, however, and expression of *cIII*, *cII*, and *cI* is low, and therefore lambda repressor is slow to accumulate, late genes such as *Q* are not repressed, so phage coat proteins and lysis proteins are produced.

    Because the growth rate of the cell has a major role in determining which pathway the lambda phage will follow, the cell does have input into the phage's decision.

16. Lambda prophage produce lambda repressor (product of the *cI* gene) while they are integrated into the *E. coli* genome. This protein, in addition to the products of "immunity genes" *rexA* and *rexB*, suppress the expression of genes necessary for the "superinfecting" phage's progression into the lytic pathway. So the presence of one prophage protects against the lytic infection by other phage.

17. a. first: *N Cro*, early: *O P cII*, mid: *cI bet exo int*, late: *rexA rexB*, end: *xis*

    b. first: *Cro N*, early: *O P cII*, mid: *Q*, late: *S R R_z*, (lysis) *Nu1 A-Z*, (head) *U-J*, (tail)

18. a. In the absense of both *cI* and *Cro*, both leftward expression and rightward expression are allowed.

    b. *Cro* protein bound to O_{R3} represses leftward expression and allows rightward expression to occur.

    c. When *cI* proteins bind at the O_{R1} or O_{R2} sites (but not at the O_{R3} site), leftward expression is allowed, and rightward expression is repressed.

    d. When either *cI* or *Cro* is bound to all three operator sites (O_{R3}, O_{R2}, and O_{R1}), both rightward *and* leftward expression are repressed.

19. a. *trpA*⁻ would reduce or eliminate tryptophan synthetase alpha activity in the cell, would probably reduce the tryptophan synthetase ($\alpha_2\beta_2$) activity, and would lead to an accumulation of InGP.

    b. *trpB*⁻ would reduce or eliminate tryptophan synthetase beta activity in the cell, would probably reduce the tryptophan synthetase ($\alpha_2\beta_2$) activity, and would lead to an accumulation of InGP.

    c. A polar mutation in *trpC*⁻ would eliminate the production of both tryptophan synthetase alpha and tryptophan synthetase beta peptides. Depending on whether the mutation eliminated *trpC* activity, reduced it, or changed its substrate range, there might be an accumulation of PRA, CdRP, or InGP.

    d. Loss of the *trpR*⁻ gene would lead to deficiency in the *trp* repressor protein. As a consequence, the initiation of transcription at the *trp* operon promoter would be the same whether there was tryptophan in the medium or not. Due to tryptophan attenuation, however, there would not necessarily be a commensurate increase in tryptophan synthetase activity or a buildup of any intermediates other than the short, prematurely terminated transcripts.

e. A mutation that generated a "super" *trp* repressor—allowing the repressor to bind whether it was bound by tryptophan or not—would be expected to cause constitutive repression of the *trp* operon. All *trp* gene expression would be reduced, as would tryptophan synthetase activity. There could be an increase in the levels of chorismate.

f. Loss of the ability of *trpO⁻* to bind the *trp* repressor would give rise to constitutive transcription initiations at the *trp* promoter, but, again because of subsequent transcription attenuation, would not necessarily give rise to increased tryptophan synthetase levels.

20. a. Because the product of *lacI*, *lac* repressor, is a *trans-acting factor*, its activity is not restricted to the molecule of DNA that encodes it. Therefore, the repressor protein will bind to both *lacO* sites and will confer allolactose-regulated repression of *lac* gene expression to both the wild-type *lacY⁺* gene that *lacI⁺* is *cis* with and the wild-type *lacZ⁺* gene that *lacI⁺* is *trans* to.

b. *lacIˢ* encodes the *lac* superrepressor that binds to functional operators and is not able to be induced to leave by allolactose. *lacIˢ* is *trans-dominant* because its product affects the expression of genes on both the DNA molecule it is on and the other DNA molecule with a *lacO* and is dominant over the wild-type gene, *lacI⁺*. In the merozygote shown in part b, no production of beta-gal over basal levels will be observed.

c. *lacI⁻ᵈ* encodes a repressor protein not able to function as an inducible DNA binding protein, but capable of interacting with other *lac* repressors to form the *lac* repressor tetramer. Because having even one subunit of the *lac* tetramer be a peptide encoded by *lacI⁻ᵈ* is enough to disrupt the repressor's ability to bind the lac operator, *lacI⁻ᵈ* is *trans*-dominant to *lacI⁺*. In the merozygote shown in part c, constitutive production of beta-galactosidase and permease would be expected.

d. *lacI⁻ᵈ* is *trans*-dominant to *lacIˢ*; therefore, the phenotype will be as in part c: constitutive production of beta-galactosidase and permease.

Eukaryotic Gene Regulation

Problems

1. Draw a schematic illustration of how transcription of the gene shown below is regulated by the transcription factor complexes when the gene is in the following transcriptional states:

   a. basal expression

   b. maximal expression

   c. silenced

   Upstream sequence

   TFA binding site   TFR binding site   Promoter

   Key: RPH   RNA polymerase holoenzyme (in this case, sufficient for transcription)
         TFA   Activating transcription factor complex
         TFR   Repressing transcription factor complex

2. In the four Southern blotting experiments shown below, DNA from hen oviduct nuclei was cut with different concentrations of DNase I, extracted, cut with a restriction enzyme, separated by gel electrophoresis, blotted to a membrane, and hybridized with the probe indicated above the schematic autoradiogram. What observations can be made about the DNase sensitivity of transcriptionally active and inactive chromatin from the Southern blot data shown below? About the regions upstream of transcriptionally active and inactive genes?

   a. Probe: beta-globin coding region

   b. Probe: beta-globin regulatory region

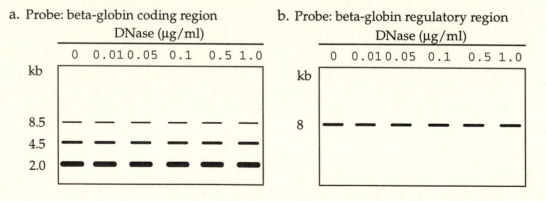

c. Probe: ovalbumin coding region   d. Probe: ovalbumin regulatory region

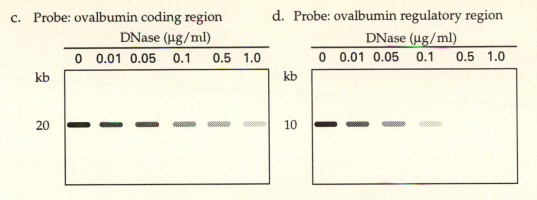

3. Histones act as general repressors of transcription. Other proteins, however, can modulate that repression. Explain a current model of how histones (acetylated and unacetylated), promoter-binding proteins, and enhancer-binding proteins interact to regulate transcription.

4. The DNA shown below corresponds to 2 kb of sequence upstream from a mammalian gene.

   a. Given the location of *MspI* sites indicated on the sequence and the results of a Southern blotting experiment in which the probe used is the entire 2 kb fragment, mark with an "M" the sites (*s) that are likely to be 5′ C^mCGG 3′ rather than 5′ CCGG 3′.

```
*          *              *          *              *          *      * * * *

|--- 200 --+------ 400 -----+-- 150 ---+------- 700 -------+-- 300 --+100-|50|30|70| →
```

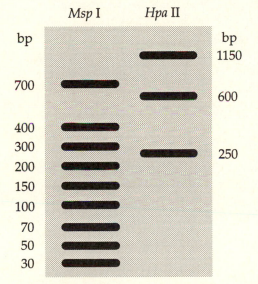

   b. Based solely on these data, would you predict that, in the cells this genomic DNA was isolated from, this gene is transcriptionally active or transcriptionally inactive?

5. The upstream activator sequence (UAS$_G$) for the *GAL1* and *GAL10* genes can be cloned upstream of a reporter gene such as the bacterial gene *lacZ* in a plasmid vector that is capable of integrating into the *Drosophila* genome when injected into early *Drosophila* embryos. The *GAL4* open reading frame can be cloned behind a weak promoter in a similar transposition vector; this vector also typically has a selectable marker that allows flies that have the *GAL4* gene integrated into their genome to be detected.

   a. Assuming that, in the absence of GAL80 in flies, GAL4 can activate transcription and that the site of integration of the *GAL4* is almost random, what you would expect to find if *Drosophila* embryos, larvae, pupae, and adults were stained to detect the presence of beta-galactosidase? Transcription factors that regulate the expression of galactose-metabolizing genes in *Drosophila* do not interact with the yeast UAS$_G$.

   b. If there existed in *Drosophila* a *GAL80* isolog that encoded a protein that could bind to the yeast GAL4 protein, what dietary supplement should the flies be given to produce the best staining?

6. Steroid and peptide hormones are released into the bloodstream and, as a result, reach nearly every cell in the body.

   a. Given how broadly hormones are dispersed in the body, how is the *specificity* of hormone action on particular target tissues accomplished?

   b. Describe the general structure of a steroid hormone.

   c. Describe the similarities and differences between heat shock response elements and steroid response elements.

7. As in all eukaryotes, gene expression in plants is regulated by transcription regulatory factors—often in association with signaling hormones—that modulate the recruitment of RNA polymerase to a given gene by binding to its promoter and enhancer regions.

   a. What are the five main classes of plant hormones?

   b. How might a gene encoding a gibberellic acid biosynthesizing enzyme be related to the Mendelian trait of plant height (TT, Tt = tall, tt = dwarf) in *Pisum sativum*?

8. Posttranscriptional control is not the same as posttranscriptional processing.

   a. How are the two terms different?

   b. What are the two major types of posttranscriptional control in eukaryotes?

   c. Give an example of how the same gene can give rise to different proteins, depending on the tissue or developmental stage in which the gene is expressed.

9. In each of the following pairs of molecules, which molecule is likely to have the longer half-life? Why?

   a. a) mRNA: 5′ (N)$_{1,500}$(A)$_{40}$ 3′

      b) mRNA: 5′ (N)$_{1,500}$(A)$_{400}$ 3′

   b. a) mRNA: 5′ m^{7}G (N)$_{1,500}$(A)$_{400}$ 3′

      b) mRNA: 5′ ATG (N)$_{1,500}$(A)$_{400}$ 3′

   c. a) vitellogenin mRNA in tissue treated with estrogen

      b) vitellogenin mRNA in tissue not treated with estrogen

   d. a) mRNA in *dcp1⁻ xrn1⁻* yeast

      b) mRNA in *dcp1⁺ xrn1⁺* yeast

  e. a)   protein: $^{+}NH_2$ Met (any amino acid)$_{400}$ COO^{-}

       b)   protein: $^{+}NH_2$ Arg (any amino acid)$_{400}$ COO^{-}

  f. a)   protein: (ubiquitin) $-^{+}NH_2$ (any amino acid)$_{400}$ COO^{-}

       b)   protein : $^{+}NH_2$ (any amino acid)$_{400}$ COO^{-}

10. Long-term gene regulation that occurs progressively and largely irreversibly during development involves changes to DNA and the cellular environment of the DNA. How did the carrot cloning experiments of Frederick Steward's laboratory in the 1950s contribute to the idea that development does not involve the tissue-specific loss of genetic material?

11. In the hemoglobinopathy beta-thalassemia, both copies of the gene encoding beta-globin are either mutated or missing altogether. How could *another* mutation in a beta-globin-like gene suppress the severe phenotype one would expect in someone lacking beta-globin genes?

12. The figure below shows a section of polytene chromosome from *Drosophila melanogaster*.

  a. What tissue was this chromosome isolated from?

  b. Label a chromomere.

  c. Label a spot on the chromosome where gene expression is very high.

  d. Label a region where ecdysone receptor (bound to ecdysone) might be expected to bind the DNA.

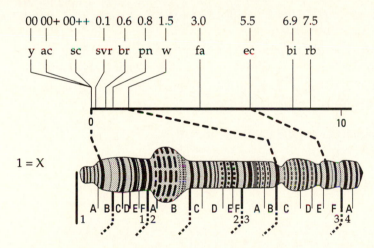

13. Genes encoding antibodies and T-cell receptors have an unusual and very important genetics.

  a. If a person has between 1 million and 100 million different versions of antibody proteins (heterotetramers)—each antibody molecule being encoded by two genes—what fraction of our genome (about 70,000 to 100,000 genes) must encode antibody peptides?

  b. What genetic processes give rise to mature mRNAs with the kind of diversity described in part a?

14. After the formation of the cellular blastoderm in *Drosophila melanogaster*, what are the two main processes that later development of body structures depends on?

15. On the 10-hour *Drosophila* embryo shown below, label the following:
    a. the segment that will generate the adult wing
    b. the segment that will give rise to the front legs
    c. the segment that will give rise to the head
    d. the segment that will give rise to the germ line (gametes)
    e. the side of the embryo that has (had) the highest level of hunchback protein
    f. the side of the embryo that has (had) the highest level of nanos protein

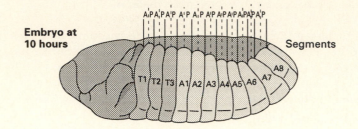

16. What general conclusions can one make about a gene in which a homeobox is found?

Challenge Problems

17. Maleness and femaleness in *Drosophila* are determined by the X:A chromosome ratio via a complex but well-characterized regulatory cascade.
    a. Which genes in the cascade produce mRNAs whose alternative splicing is essential to the sex determination pathway? How is each gene's alternative splicing regulated?
    b. What genes encode the transcription factors mainly responsible for repressing male-differentiation genes? For repressing female-specific genes?

18. Classify and predict the phenotypes of the following mutations:
    a. *nanos*⁻
    b. *odd-skipped*⁻
    c. *hunchback*⁻
    d. *torso*⁻

Team Problems

19. In 1997, Ian Wilmut's laboratory published a paper describing the cloning of a sheep named Dolly.
    a. Describe the process of cloning Dolly.
    b. In the process, what was learned about the developmental potential of a mammalian mammary gland cell?
    c. What do you think some of the ethical complexities of cloning a human would be?

20. Describe the levels of regulation of gene expression in eukaryotes, note which are important in prokaryotes, and give examples of eukaryotic genes controlled at each level where possible.

Solutions

1. a.

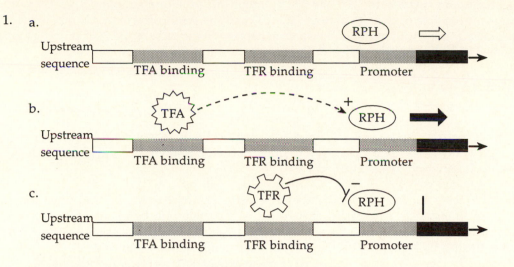

2. Figures a and b show that a gene that is *not* transcriptionally active (beta-globin, in hen oviduct cells) is resistant to DNase digesting, presumably due to its compacted state. Increasing quantities of DNase do not cut either the coding or regulatory regions of beta-globin. Figure c shows that ovalbumin is sensitive to digestion by DNase. As the concentration of enzyme increases, the genomic DNA fragment containing the ovalbumin coding region diminishes in abundance, reflecting its accessibility to the DNase. Figure d illustrates the phenomenon of hypersensitivity of the regulatory region of transcriptionally active genes. Upstream regulatory regions are *hypersensitive sites* in the chromatin because they are even more "open" than the coding sequence that follows them.

3. In one model, enhancer-binding proteins destabilized the histones bound to the promoter region, allowing promoter-binding proteins (such as TATA-binding factors) to interact with the promoter where previously histones bound preferentially. Histones in the vicinity of an active area of transcription are acetylated on lysines on the outside face of nucleosomes. The acetylated histones are less able to form higher-order structures (more condensed chromatin), making the repressive action of the histones easier to overcome.

4. a. To solve this problem, recall that both *Msp*I and *Hpa*II cut DNA at the same site: (5′ CCGG 3′), but that methylated cytosines (^mC) within the *Hpa*II recognition prevent *Hpa*II cleavage. If none of the cytosines were methylated, the *Hpa*II and *Msp*I restriction patterns would be identical. Referring to the *Hpa*II digest lane on the Southern blot, however, we can see the fewer and larger bands that reflect the loss of *Hpa*II sites due to methylation.

    To determine which sites have been lost, choose a band (say, the 1,150 bp band) in the *Hpa*II digest lane, and look at the *Msp*I restriction map of the fragment given in the problem. Which fragments would have to remain together (i.e., *not* be separated by restriction digestion) to obtain a 1,150 bp band? We see there are two possibilities. The 1,150 bp fragment = (150 + 700 + 300) *or* (700 + 300 + 100 + 50). The second possibility can be eliminated, however, by seeing that keeping those fragments together would leave a 750 bp fragment and a 150 bp fragment, neither of which is seen on the Southern blot's *Hpa*II lane.

The possibility that the 1,150 bp band is a product of failure to cut between the 150 and 700 bp fragments and between the 700 and 300 bp fragments (due to methylation) leaves a 600 bp fragment possible (if the site between the 200 and the 400 bp fragments is methylated) and a 250 bp fragment (if many of the sites on the right-hand side of the 2 kb fragment are methylated). The best interpretation of the data in the Southern blot is shown below.

```
*      M           *        M              M        *    M M M *

|--- 200 ---|------ 400 ------|--- 150 ---|-------- 700 --------|--- 300 ---|-100-|50|30|70|→
```

    b.   Such ample methylation is suggestive of a transcriptionally quiescent gene because methylation is generally inversely proportional to transcription. It is likely that the gene is inactive in the cells from which it was purified.

5. a.   As shown in the schematic below, GAL4 will be made when the transgene encoding it integrates near an enhancer. The tissues in which GAL4 protein will be made depend on the developmental and tissue specificity of the enhancer the *GAL4* gene integrates next to. If, for example, the nearby enhancer is active only in the gut and salivary glands and only during embryogenesis, then GAL4 protein will be made in above-background quantities only in the gut and salivary glands and only during embryogenesis. The GAL4 will also be available to bind the UAS$_G$-*lacZ* fusion only in those tissues and developmental stages, so if embryos are stained for beta-galactosidase activity, it will be detected only in the gut and salivary glands.

The general usefulness of this procedure is great. Not only does one detect (or "trap") enhancer regions in the *Drosophila* genome, but one knows which tissues the enhancer is active in and when. Furthermore, one can now cross the strain of flies that contain the particular location of *GAL4* integration to flies containing UAS$_G$-*gene X* fusions and can examine such important questions as what happens to the organism when protein X is made in that particular tissue at that particular time.

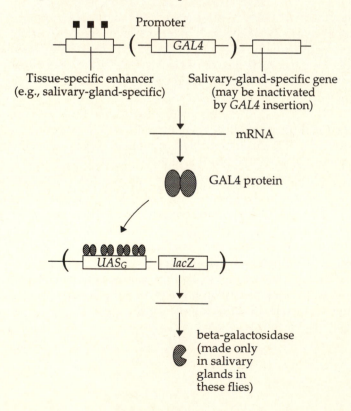

b. Flies would need to be fed galactose if they had a functional equivalent of the GAL80 protein that could inhibit the GAL4 protein in the absence of galactose metabolites. The success of the experiment as described would depend on *GAL4* to encode a protein active in the absence of GAL80 or to clone *GAL80* along with the *GAL4*.

6. a. The tissue specificity of hormones is determined by which cells in the body have receptors for the steroid on their cell surface (for peptide hormones) or in their cytosol or nuclei (for steroid hormones). The presence of the right combination of other regulatory proteins is also important in determining which tissues will respond to a given hormone and under which conditions.

    b. Steroid hormones all have a four-ring structure with differences in the particular side groups attached to the rings. These side groups are largely responsible for determining the particular receptor the steroid will bind and the particular physiological effect it will have. The female hormone progesterone differs from the male hormone testosterone only in that the hydroxyl group of testosterone is replaced by an acetyl group in progesterone.

    c. Both heat shock response elements and steroid response elements tend to have twofold rotational symmetry and typically occur in multiple copies in the regulatory sequence of the genes they regulate. Heat shock response elements, however, are located within the promoter region of the genes they regulate, and steroid response elements tend to be in enhancer regions.

7. a. The five main plant hormones known are ethylene, abscisic acid, auxins, cytokinins, and gibberellins.

    b. Because gibberellic acid can be added to a dwarf (*tt*) plant to make it grow taller, it is a reasonable hypothesis that, in these dwarf plants, a gene that encodes a protein upstream (biochemically) from gibberellic acid might be defective in the dwarf plants. A gene encoding an enzyme that synthesizes gibberellic acid is one good candidate for the gene defective in dwarf plants.

8. a. Posttranscriptional control refers to the regulation system that determines which of several possible posttranscriptional processing steps will actually be carried out for a particular gene. Small nuclear ribonucleoprotein particles (snRNPs) are likely to play a role in both RNA processing and regulation of that processing. Other proteins, such as cleavage stimulating factor (CstF), are important primarily for regulation of processing.

    b. For mRNA, choice of poly(A) site and choice of splice site are the two major steps in RNA processing that are regulated events.

    c. The μ gene that encodes the heavy chain of IgM molecules produces a transcript that early in development produces a transcript that is cut at a poly(A) addition site downstream of the exon that encodes a transmembrane domain, so the protein produced is associated with the cell membrane. Later in life a poly(A) addition site upstream of the transmembrane domain exon is used, giving rise to secreted IgM proteins that don't have the C-terminal transmembrane domain.

9. a. **b)** mRNA: $5'\ (N)_{1,500}(A)_{400}\ 3'$ because of the longer poly(A) tail. Longer poly(A) tails protect mRNAs from degradation better than short tails. mRNAs bound by poly(A) binding protein (PAB) are stable and protected from decapping and endonucleases. When deadenylation shortens the tail to the point where too few binding sites for PAB remain, the mRNA is decapped and degraded shortly thereafter.

    b. **a)** mRNA: $5'\ m^7G\ (N)_{1,500}(A)_{400}\ 3'$ because the $5'$ cap is very important for protecting the mRNA from degradation by $5'$ to $3'$ exonucleases.

c. **a)** vitellogenin mRNA in tissue treated with estrogen because estrogen is a regulatory signal, or effector molecule, that promotes stability of vitellogenin mRNA, which encodes the protein precursor to egg yolk. In the presence of estrogen, frog liver vitellogenin mRNA has an approximate half-life of 500 hours; in the absence of estrogen, the mRNA half-life is only 16 hours.

d. **a)** mRNA in *dcp1⁻ xrn1⁻* yeast would be expected to have a longer half-life because both a major decapping enzyme and a major 5′ to 3′ exonuclease are deficient because the genes encoding them are defective in *dcp1⁻ xrn1⁻* yeast.

e. **a)** protein: $^+NH_2$ Met (any amino acid)$_{400}$ COO⁻ would be more stable because of the N-end rule: proteins having an N-terminal methionine, proline, valine, glycine, threonine, serine, alanine, or cysteine have a greater than 400× longer half-life in yeast than proteins with an N-terminal arginine, lysine, phenylalanine, leucine, or tryptophan.

f. **b)** protein: $^+NH_2$ (any amino acid)$_{400}$ COO⁻ would have a longer half-life because ubiquitin (conjugated with the other protein) is used to mark proteins for degradation.

10. Because Steward could grow an entire carrot (leaf, stem, root, epithelium, phloem, and xylem) from a single, differentiated phloem cell by adding the cell to sterile nutrient medium and incubating it for several weeks, it could not be that genes essential for development of any tissue were lost during phloem development. If they had been, the cells would not have been totipotent (able to give rise to the whole organism).

11. Sometimes there is a mutation in the enhancer that regulates the developmental regulation of either γG-globin or γA-globin. These fetal peptides are similar to the beta-globin protein adults usually have and can form heterotetramers ($\alpha_2\gamma A_2$ or $\alpha_2\gamma G_2$) capable of binding iron and carrying oxygen. If the mutation in the enhancer has the effect of prolonging production of the γG-globin or γA-globin peptides, then much of the beta-thalassemia phenotype is suppressed.

12. a.   Salivary gland polytene chromosome.

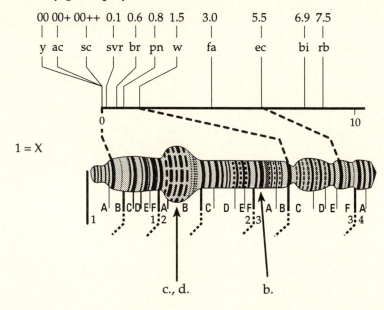

13. a. There are five different types of antibody proteins: IgA, IgD, IgE, IgG, and IgM. Each is composed of a characteristic pair of heavy chains (encoded by the α, δ, ε, γ, and μ genes, respectively) and a pair of light chains (encoded by one of two genes, κ or λ). These seven genes could encode 10 proteins (5 [number of heavy chain genes] × 2 [number of light chain genes]) if they were not rearranged, spliced, or modified in any way. Yet these genes alone encode for the vast majority of the antibodies in the organism. 10/100,000 = 1/10,000 of the genome is all that is required.

    b. Generation of antibody diversity depends on two main genetic sources of variation. One of them, alternative splicing, occurs at many gene loci in the genome. In production of kappa light chains, alternative splicing governs the choice of J (joining) exons that will be found in the mature mRNA. In production of the heavy chains, alternative splicing determines which C (constant) exon is chosen from the several possible.

    The second source of genetic variation in antibody genes appears to be *specific* to antibody genes (though similar processes occur in yeast and other organisms). Somatic recombination occurs at the DNA level with rearrangement—and actual *loss*—of genetic material. At the kappa light chain locus, V (variable) exons are brought together with a J (joining) exon. At the heavy chain locus, V exons are brought, along with an L (leader) sequence, to be adjacent to a D (diversity) region while the intervening DNA is excised and degraded. Additionally, the L V D module is moved next to one of the J regions by the same process, giving rise to a particular L V D J combination. Finally, the L V D J module is brought in apposition to a particular C (constant) region to generate the mature *gene* in a particular antibody-producing cell lineage. Splicing following transcription then generates the final mRNA, as described previously.

14. The two most important processes for patterning the early *Drosophila* embryo are (1) molecular gradients (of mRNA, protein, or active forms of protein) along the anterior-posterior axis and the dorsal-ventral axis of the *Drosophila* embryo and (2) division of the embryo into segments that give rise to the segmental structure in the adult organism.

15.

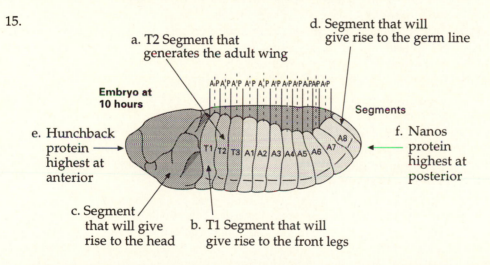

a. T2 Segment that generates the adult wing

d. Segment that will give rise to the germ line

Embryo at 10 hours

Segments

e. Hunchback protein highest at anterior

f. Nanos protein highest at posterior

c. Segment that will give rise to the head

b. T1 Segment that will give rise to the front legs

16. Homeoboxes are characteristic to (and define) homeotic genes. Homeobox genes tend to be clustered in complexes (which are themselves likely to have been relics of ancient duplication events). They all encode DNA-binding proteins with helix-turn-helix motifs. For the most part, they regulate transcription in a spatially restricted (generally tissue-specific) manner. Homeobox-containing genes (*Hox* genes) specify body plans in both invertebrates and vertebrates.

17. a. *Sex lethal (Sxl)* is spliced in females to remove a premature stop codon, but not in males. Therefore, *Drosophila* females have functional sex lethal protein, but males do not. Splicing of *Sxl* mRNA is regulated by factors responsive to X:A ratio.

    Transformer (tra) is also alternatively spliced in males and females. Sex lethal protein regulates which of two alternative splicing pathways is followed.

    Doublesex (dsx) mRNA splicing is also regulated. Transformer protein, present in females only, joins transformer-2 protein to give rise to female-specific *dsx* transcripts. Male-specific versions of *dsx* transcripts and proteins are made without transformer protein.

    b. In females, the *intersex* gene encodes the intersex protein, which, with the female-specific product of the *doublesex* gene, represses the expression of male-differentiation genes, giving rise to female somatic cells. In males, the male-specific product of the *doublesex* gene represses the expression of female-differentiation genes, giving rise to male somatic cells.

18. All of these are maternal effect mutations. All are lethal.

    a. *nanos* is a gap gene required for the generation of posterior segments. Females homozygous for mutations in *nanos* produce eggs that become embryos without abdomens.

    b. *odd-skipped* is a pair-rule gene complementary to *even-skipped*. Females homozygous for *odd-skipped* mutations produce eggs that develop into embryos missing every odd segment.

    c. *hunchback* is a gap gene that, when homozygous in females, causes eggs produced to become embryos missing their anterior (thoracic) segments.

    d. *torso* is a gap gene specifying terminal structures in the *Drosophila* embryo. Females homozygous for *torso* mutations produce eggs that become embryos lacking terminal anterior structures, having terminal posterior structures on both anterior and posterior ends.

19. a. Dolly was the product of an enucleated (nulliploid) egg fused with a mammary gland cell (diploid). Prior to fusion, the mammary gland cell had been grown in tissue culture and induced to enter a quiescent state by serum starvation in the growth medium. Once the donor cell was in G_0 of the cell cycle, presumably it was able to be "reset" by cytoplasm of the nucleus-free egg it joined. The fusion cell divided mitotically over the course of 6 days. The resulting embryo was implanted into a recipient ewe, and Dolly was born, almost identical genetically to the sheep from whose mammary cell she had been cloned.

    b. What is surprising is that the mammary gland cell nucleus did not retain its restricted mammary gland cell state and generate only mammary gland cells or partially retain its restricted state and generate a nonviable cell mass. What only weeks before had been directing the activities of a differentiated cell was now the nucleus governing the development of an entire organism. The revolutionary conclusion from the work is that the nuclei of mammalian cells are *totipotent* and can, under the right circumstances, be induced to switch from a differentiated cell program to an embryonic development program.

    c. There are many, including the following: (1) What happens to all the embryos that are generated in the process that are not chosen for implantation? (2) What effects might there be for the child—possibly including shortened lifespan, increased susceptibility to cancer, and many others unanticipated? (3) Who is responsible for the child: only the person cloned, their partner, too, others? (4) If the justification for the cloning is to generate tissue (bone marrow, liver, etc.) for the person cloned, what are the child's rights? (5) What if, instead of generating a whole organism, it was possible to "make" a mass of tissue of a particular type? Would that be different? There are, of course, many other questions that the prospect of human cloning brings up.

20. 1. Somatic recombination. *Genes* that encode peptides that will be part of antibodies or the T-cell receptor are processed to their mature form in B-cell precursors and T-cell precursors, respectively. In other cell types, this rearrangement does not occur.

2. Transcriptional control is the most common level of regulation. Transcription factors binding to the promoter and enhancer of the *GAL1* and *GAL10* gene in yeast, for example, regulate the transcription of DNA and, thereby, the production of galactose-metabolizing proteins. Transcriptional control occurs in prokaryotes, though it is often in the context of an operon, where several genes are transcribed together from a common promoter.

3. Precursor RNA processing. The *sex-lethal* RNA, for example, is spliced differently, depending on whether the X chromosome to autosome ratio (X:A) is 1 (in females) or 0.5 (in males). Prokaryotes do not as a rule process their RNA by splicing.

4. mRNA transport from nucleus. When a long poly(A) tail is indicated by a particular polyadenylation signal, the mRNA is more stable and more likely to escape the nucleus. Prokaryotes, with neither nucleus nor significant addition of poly(A) tails, do not have this level of regulation.

5. Translation of mRNA. The sequence surrounding the upstream AUG codons in the mRNA determine not only which start of translation will be used by ribosomes, but the rate of mRNA translation. This sequence is called the Kozak sequence in eukaryotes and Shine-Dalgarno in prokaryotes. Both eukaryotes and prokaryotes regulate gene expression at this level.

6. Degradation of mRNA. Casein mRNAs (that encode the major milk protein) are stabilized in the presence of prolactin, the hormone that induces milk production. The half-life of prolactin is 19 times longer in the presence of prolactin than in prolactin's absence.

7. Degradation of protein. Lens proteins of the vertebrate eye can last the lifetime of the organism, but steroid receptor proteins are short-lived. Some of these differences in protein half-life depend on the amino acid at the N-terminus of the protein. Though methionine is usually the first amino acid of the protein's N-terminus, proteins are frequently trimmed at their N-termini ends to remove signal sequences. In the process, an amino acid such as phenylalanine or tryptophan could be at the new N-terminus, reducing the half-life of the protein dramatically.

22
Mutation

Problems

1. Comment as fully as you can on the nature of each of the following mutations based on the data provided.

   a. Homozygous mutation 1a has a phenotype that is half as severe as a homozygous null mutation of the same locus.

   b. Mutation 1b when heterozygous with the wild-type allele yields a more severe phenotype than an organism heterozygous for a null mutation.

   c. Mutation 1c is on the X chromosome; men who have it tend to have a wide variety of problems that cannot be explained by simple pleiotropy.

   d. Homozygotes for mutation 1d1 have congenital blindness. Homozygotes for mutation 1d2 are also congenitally blind. All the children of a man and a woman of these two genotypes, however, can see.

   e. Mutation 1e tends to be observed as a somatic mutation. A protein that is normally expressed at low levels is found at constitutively high levels and is associated with unrestrained growth of the cell in which the mutation is found.

   f. When mutation 1f is maternally inherited, the affliction is characterized by inappropriate laughter and jerky movements. When the same mutation, 1f, is paternally inherited, the affliction is characterized by weakness and obesity.

   g. For mutation 1g, there is the normal amount of mRNA produced, but a reduced amount of a protein that is smaller than usual.

   h. Tumor cells with a single copy of mutation 1h contain a transmembrane protein that now possesses a cytoplasmic kinase domain that is not part of the normal transmembrane protein.

   i. Though the predicted amino acid change in the protein encoded by mutation 1i is conservative and doesn't appear to correspond to a crucial position in the protein's active site or structural motifs, the mRNA is unusually large, and the protein is unusually small (or absent).

2. *E. coli* are grown in minimal medium containing lactose as the only carbon source until their density is 10^4 cells/milliliter. One milliliter of these cells is then put into 20 milliliters of rich growth medium containing glucose instead of lactose, but now also containing acridine orange, an intercalating mutagen. These cells are cultured for 2 hours, the length of time it takes for *E. coli* to complete four cell divisions at 37°C. Cells are then plated on agar plates containing X-gal, a galactose analog that turns blue in the presence of β-galactosidase, and isopropyl-β-D-thiogalactoside (IPTG), an inducer of the *lac* operon. Fifty-eight white colonies are observed and streaked on agar/X-gal/IPTG plates to isolate pure clones of each. The *lacZ* gene in each of the 58 clones is then sequenced, leading to the discovery that there are 20 independent mutations.

   a. What is the *mutation frequency* for the *lacZ* gene at the end of the experiment?

   b. What is the *mutation rate* for the *lacZ* gene?

   c. What kind of mutations would you expect to be most frequent among the 20?

3. The sequence below corresponds to the first 15 bases of the open reading frame of the *E. coli cysS* gene. For each of the following mutant sequences, determine the sequence encoded by the 15 bases, identify whether the mutation is a transition, transversion, insertion, or deletion, and identify whether the mutation is a missense, neutral, nonsense, silent, or frame-shift mutation.

$$5'\ \text{ATG CTA AAA ATC TTC} \ldots$$
$$\text{H}_3\text{N}^+\ \ \text{M}\ \ \ \text{L}\ \ \ \ \text{K}\ \ \ \ \text{I}\ \ \ \ \text{F}$$

a.   5' ATG CTA AAG ATC TTC ...

b.   5' ATG CTA TAA ATC TTC ...

c.   5' ATG CTA AAA AAT CTT ...

d.   5' ATG CGA AAA ATC TTC ...

e.   5' ATG CTA AGA ATC TTC ...

4. Peptides α-globin and β-globin interact as a heterotetramer, with two of each peptide and four heme groups required to make one hemoglobin A protein. In the figure below, a portion of each α-globin peptide is shown with some of the amino acids essential for the electrostatic interaction between α-globins connected by lines. Amino acids are indicated by their one-letter code.

```
      ...  LDKFLASVSTVLTSKYR          COO⁻
                |             |
      ⁻OOC     RYKSTLVTSVSALFKDL ...
```

```
413             428            443            458
ctg gac aag ttc ctg gct tct gtg agc acc gtg ctg acc tcc aaa tac cgt taa
 L   D   K   F   L   A   S   V   S   T   V   L   T   S   K   Y   R   .
126             131            136            141
```

The sequence of bases in the α-globin gene that encodes the portion of the peptide shown here is also given. Note that the peptides interact over the same section in both cases, but have opposite polarity.

Note that in humans, there are two loci encoding α-globin, so that in any given diploid individual, there will be four alleles encoding α-globin.

Nomenclature for base-substitution mutations is base # (A → B) where A is the base at the base number indicated and B is the base it has been mutated to. So 432 (G → A) indicates a change of the guanine at position 432 to an adenine.

Nomenclature for mutations that cause a change of amino acid in a protein is A(amino acid #)B, where A is the amino acid normally found at the site indicated, and B is the amino acid found there in the mutant being described. Y36F indicates that amino acid 36 is tyrosine (Y) in the original protein, but phenylalanine (F) in the mutant protein. X is used to designate a stop codon.

a. Using proper nomenclature, what mutation in the α-globin sequence shown would cause an *intragenically suppressible forward mutation*?

b. Again, using proper nomenclature, what is a mutation that would be a *partial reversion* of the mutation you describe in part a?

c. Would there be a *full reversion* of the mutation in part a?

d. Give an *intragenic suppressor* of the mutation in part a.

e. Give an *intergenic suppressor* of the mutation in part a.

5. What is the difference between spontaneous mutations and induced mutations? Give examples of each.

6. How is the wobble pairing hypothesis of base pair mismatch formation different from the rare tautomeric forms hypothesis?

7. Illustrate how the following mutation could occur:

     5′ CTATCTATCTCTATCTCTATCTCTATCTCTATCTCTATATATC 3′

     3′ GATAGATAGAGATAGAGATAGAGATAGAGATAGAGATATATAG 5′

                      ↓

     5′ CTATCTATCTCTATCTCTATCTCTATCTCTATCTCTATCTCTATATATC 3′

     3′ GATAGATAGAGATAGAGATAGAGATAGAGATAGAGATAGAGATATATAG 5′

8. Deamination is one of the two most common chemical changes that lead to spontaneous mutations.
   a. Draw the chemical structure of deoxy-5-methylcytosine monophosphate.
   b. What base would usually pair with 5-methylcytosine?
   c. Draw the deamination product of deoxy-5-methylcytosine monophosphate.
   d. What base would usually pair with that deamination product?
   e. What kind of base substitution mutation (transition or transversion) accompanies deamination of 5-methylcytosine?
   f. What codons can be converted to stop codons by mutations that result from 5-methylcytosine deamination?
   g. Why is a 5-methylcytosine more likely than a cytosine to be a "mutation hotspot?"

9. How is DNA repaired following a depurination event?

10. What is the basis for the mutagenicity of ultraviolet (UV) light?

11. 2-aminopurine (2-AP) can revert 5-bromouridine (5BU)-induced mutations and vice versa. Hydroxylamine (NH_2OH), however, can revert only half of the mutations caused by 2-AP or 5BU. Explain.

12. Some of your *E. coli lacZ* mutants make a truncated β-galactosidase protein as detected by loading extract onto a polyacrylamide gel for resolving proteins by electrophoresis and blotting the proteins onto a membrane for detection with anti-β-galactosidase antibodies. What would be a reasonable mutagen to use in an attempt to revert these particular mutants?

13. The Ames test has been very important in establishing mutagenicity of chemicals of previously unknown toxicity. What mutations are already present in the test organism (*Salmonella typhimurium*) before chemicals to be tested are added? What purpose does each of these preexisting mutations serve?

14. What are the main DNA repair mechanisms present in *E. coli*? What genes and proteins are involved?

15. Illustrate the process of mismatch correction mediated by the MutS, MutL, and MutH proteins in *E. coli*.

16. Xeroderma pigmentosum is an autosomal recessive disease that gives rise to symptoms such as extreme sensitivity to sunlight (freckling, melanomas, etc.) and early lethality. What bacterial DNA repair process is most likely homologous to the process disrupted in those with xeroderma pigmentosum?

Challenge Problems

17. Describe in detail how you could test the hypothesis that histidine 5 (His5) encoded by the plasmid-borne *Amp^r* gene shown below (in part) is *essential* for function—without performing an extensive random mutagenesis.

```
5′ ... ATG AGT ATT CAA CAT TTC CGT GTC GCC CTT ATT ...
        M   S   I   Q   H   F   R   V   A   L   I
```

18. Propose a screen for obtaining temperature-sensitive mutations in genes required for galactose metabolism in yeast.

Team Problems

19. Describe the process of making a knockout mutation of the β-globin gene in mice.

20. What are some of the possible reasons why all organisms have an endogenous mutation rate greater than 0?

Solutions

1. a. Allele 1a is a partial *(hypomorphic) loss of function* mutation. There is either less protein or the protein is less functional than the normal gene product.

    b. Allele 1b is called a *(dominant negative)* mutation, somehow antagonizing the wild-type gene/gene product, interfering with its normal functioning.

    c. Mutation 1c is probably a chromosomal mutation consisting of a deletion of many genes. The result is a *contiguous gene syndrome*, in which symptoms arise from the mutation of several different, contiguous genes.

    d. Mutations 1d1 and 1d2 are in different gene loci so that *intergenic complementation* is seen, the children all being heterozygotes M/+ at both loci. Less likely, but still possible, this could be an example of *intragenic complementation*, where the mutations are in the same gene, but the two together are capable of producing the necessary "activity" for a normal phenotype.

    e. Mutation 1e is most likely an activating mutation in an *oncogene*, that is, the mutation that alters a gene product that normally regulates cell division. In this case, it appears the mutation is a regulatory sequence that directs the quantity of oncogene product to synthesize or affects the product's rate of degradation.

    f. Mutation 1f is in an *imprinted* region of the chromosome. In the mother's gametes, it is methylated one way, in the father's, another. In this case, both sexes appear to require the gene product.

g. Allele 1g probably has a *nonsense mutation*. That is, though the RNA is of the expected size and abundance, the mutation has generated a new stop codon in the coding sequence that gives rise to a truncated protein. Other possibilities include a *frameshift mutation* that leads to a premature stop or the mutational *activation of a cryptic proteolytic cleavage signal*.

h. Mutation 1h is probably a *gene fusion* that generates a *chimeric protein*. The protein probably has new functions that neither of its parent molecules had. Given that the heterozygous phenotype would probably not resemble the loss of function of either parent gene, this would be classified as a *neomorphic mutation*.

i. Mutation 1i may be at an exon-intron boundary such that the transcript is not properly spliced. If an intron were retained, the processed mutant transcript would be longer than the normal mature transcript. The protein would most likely be truncated due to stop codons within the kept intron.

2. a. mutation frequency = number of mutations/total number of cells

$$= 58 \text{ mutations}/[1.6 \times 10^5 \text{ cells*}]$$

$$= 3.6 \times 10^{-4} \text{ mutations/cell}$$

*total number of cells = 10,000 + 10,000 + 20,000 + 40,000 + 80,000

0' (start) 30' (div) 60' (div) 90' (div) 120' (div)

b. mutation rate = number of mutation events/gene/cell division

$$= \frac{20 \text{ } lacZ \text{ mutations}}{[(1 \times 10^4) + (2 \times 10^4) + (4 \times 10^4) + (8 \times 10^4) \text{ cell divisions}]}$$

$$= 1.33 \times 10^{-4} \text{ mutations per } lacZ \text{ gene per cell division}$$

c. Small insertions and deletions (causing frameshifts) should be the most prevalent kind of mutation because acridine orange is an intercalating agent. Intercalating agents wedge between base pairs in the double helix, causing distortions that lead DNA polymerase to make insertion and deletion errors.

3. a. M L K I F ...   transition, silent mutation
    b. M L .   transversion, nonsense mutation
    c. M L K N L ...   insertion, frameshift mutation
    d. M R K I F ...   transversion, missense mutation
    e. M L R I F ...   transition, neutral mutation

4. a. Many are possible, each with a different "ease" of partial reversion or suppression. Because the problem discusses only one interaction (and its reciprocal) between α-globin peptides, the one between Asp127 and Arg142, bases encoding these amino acids are the most likely to give a mutant phenotype when altered. Because both aspartate and arginine are charged amino acids, finding a base change that results in a codon for a nonpolar or *oppositely* charged amino acid is not difficult. Arginine here is encoded by CGT. A transversion of the G to T gives CTT, which encodes for the nonpolar amino acid leucine. That gene mutation is described as 462 (G → T) and the protein mutation as R142L. This is one answer to this question; others can be obtained by following a similar logic.

b. A partial reversion of the mutation in part a would be transversion 462 (T → A). This accomplishes the return of a basic amino acid, histidine, to position 142 (L142H). The reversion is partial because arginine and histidine differ in shape and in precise chemistry, factors that may alter the stability of the hemoglobin tetramer.

c. A full reversion of the original mutation would be the back mutation, 462 (T → G). This change would restore codon 142 to one that encodes arginine. Other mutations that would return the codon to one encoding arginine require more than a single base change to accomplish.

d. An intragenic suppressor of R142L *that is not simply a reversion* is somewhat difficult to come up with for this particular example without envisioning a somewhat complicated mutation process or making assumptions about the structural plasticity of the hemoglobin A protein. One solution to this problem, however, could be the mutation 458 (T → C), which would change codon 141 from TAC to CAC, which would produce a protein with histidine at position 141 rather than tyrosine (Y141H). Though the stereochemistry might not be good, and this leaves a potentially destabilizing nonpolar residue at the terminal, it is possible that putting a positively charged amino acid (histidine) next to the nonpolar terminal leucine would restore the electrostatic interaction with the aspartate of the other peptide. This is an *intragenic* second-site suppressor mutation because the base changed to restore normal functioning of the gene product is not the same one as caused the first mutation.

Another possible intragenic suppressor mutation might be a 6-base-pair insertion just following base 452. This would effectively push the lysine that is now at position 140 to position 142, allowing it to interact with Asp127. It would be important that the two new codons do not introduce amino acids that destabilize the tetramer. It would also be important that having two additional amino acids in the peptide (including a terminal nonpolar amino acid) not destabilize the tetramer.

e. An *intergenic* suppressor mutation of R142L might be mutation in the other α-globin locus. Recall that there are two completely different loci, with a total of four α-globin alleles. One suppressor might be a transversion mutation of 417 (A → T), converting codon 127 from one encoding aspartate (GAC) to one encoding valine (GTC). Like leucine, the amino acid found in the mutant α-globin gene, valine is nonpolar. These two amino acids could, conceivably, give a more stable heterodimer than one would get without the suppressor mutation.

5. Spontaneous mutations are a result of errors in DNA replication and of spontaneous changes in the DNA (due to endogenous transposable elements, oxidation, incidental radiation, etc.). Induced mutations are those caused intentionally by a scientist by use of exogenous chemical mutagens, radiation (UV, X rays), or experimentally induced mobilization of transposons.

6. One hypothesis for how base mismatches occur is that bases sometimes (rarely) take on an alternate chemical form, becoming the tautomer of the usual Watson-Crick form of the base. These tautomers have different base-pairing properties, allowing G to pair with T and C to pair with A when, in each case, one member of the pair is in its alternative form.

An alternative hypothesis for how base mismatches occur is via "wobble base pairing." In this model, DNA takes an unusual conformation, changing the spatial arrangements of atoms in the chain of nucleotides. During DNA replication or repair, non-Watson-Crick base pairs are formed because the base being added cannot form a normal Watson-Crick base pair with the DNA in that configuration.

Evidence supporting the "wobble base pairing" hypothesis is mainly of two sorts: (1) alternative (tautomeric) forms of bases may be *much* less common than previously thought, and (2) mismatched base pairs are observed in their normal forms by X-ray crystallography and NMR analysis.

7. Slipped strand mispairing is often responsible for the generation of small insertions or deletions. The schematic below illustrates how a newly synthesized strand can lift up from the template and reanneal with the wrong bases, producing an out-of-register looped-out structure, which produces a deletion or an insertion alone or more repeats when DNA replication continues.

```
5' CTATC TATCTC TATCTC TATCTC TATCTC TATCTC TATATATC 3'
3' GATAG ATAGAG ATAGAG ATAGAG ATAGAG ATAGAG ATATATAG 5'
                  T C
              A       T    ↓ mid-first round of replication
                  T C
5' CTATC TATCTC TATCTC TATCTC TATCTC TATCTC TATCTC ...        3'
3' GATAG ATAGAG ATAGAG ATAGAG ATAGAG ATAGAG ATAGAG ATATA TAG 5'
                           ↓ after second round of replication


5' CTATC TATCTC TATCTC TATCTC TATCTC TATCTC TATCTC TATATATC 3'
3' GATAG ATAGAG ATAGAG ATAGAG ATAGAG ATAGAG ATAGAG ATATATAG 5'
        6 repeats
                                    and
5' CTATC TATCTC TATCTC TATCTC TATCTC TATCTC TATATATC 3'
3' GATAG ATAGAG ATAGAG ATAGAG ATAGAG ATAGAG ATATATAG 5'
        5 repeats
```

8. a.

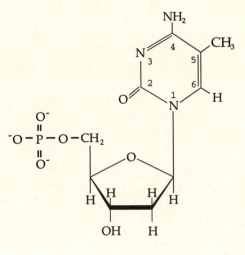

Deoxy–5–methylcytosine monophosphate

b. Guanine normally pairs with 5-methylcytosine.

c.

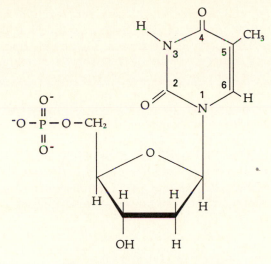

Deoxythymidine monophosphate

d. Adenine normally pairs with thymine.

e. Deamination of 5-methylcytosine leads to transitions (always G → A and C → T).

f. TGG (tryptophan-encoding) codons can be converted to either TGA or TAG (or TAA, in the case of the rare double deamination) by the transition of G → A at either the second or third position in the codon.

g. The deamination of cytosine produces uracil. Uracil is recognized by the DNA repair machinery as *not* a DNA base and can therefore be removed and repaired.

9. After depurination, one of two things happens. The area of depurination can be repaired by removing some of the bases on the depurinated side and replacing them using the nondepurinated side as template. Alternatively, the area of depurination may *not* be repaired. If this is the case, when the DNA replicates, a base is chosen at random opposite the site of depurination.

10. Purines and pyrimidines absorb UV light maximally in the 255–260 nm wavelength range. Not only is this feature good for determining concentrations of DNA, but it also can cause photochemical bonds. The most common is the thymine dimer, in which consecutive thymines in a DNA strand are cross-linked by UV radiation. Many thymine dimers are repaired by the SOS pathway. Photolyase, with light at a wavelength of 355 nm, catalyzes the conversion of thymine dimer back to individual thymines. If there are too many thymine dimers, however, the cell may enter a programmed cell death pathway.

11. Although 2-AP and 5BU are capable of causing transitions of either type (GC → AT or AT → GC) and hence can revert any mutation they cause, hydroxylamine induces only CG → TA transitions. It will therefore be limited in which transitions it can revert.

12. These truncated proteins could be caused by nonsense mutations in the open reading frame of *lacZ* or by frameshift mutations that shift the reading frame to one with a stop codon following shortly after the site of frameshift. Nonsense mutations can be reverted by 2-aminopurine or 5-bromouridine. Frameshifts are most easily reverted with intercalating agents such as proflavin, acridine orange, or ethidium bromide. The choice of mutagen should take into account the mutagen used in the initial mutagenesis.

13. Histidine synthesis is one preexisting mutation. Histidine auxotrophy allows one to select for *his*⁺ revertants. The mixture includes frameshift, transition, and transversion mutant strains to detect different kinds of mutagens. Another is lipopolysaccharide coat. Lacking the lipopolysaccharide coat, chemicals being tested have greater access to bacteria. A third preexisting mutation is SOS repair. More of the DNA damage caused by the test chemical will be converted to a mutation rather than being repaired.

14. Repair mechanisms include (1) 3' to 5' proofreading (encoded by *mutD*⁺) of DNA polymerase III, (2) photoreactivation (mediated by photolyase encoded by the *phr* gene), (3) alkyl transfer (catalyzed by enzymes such as O⁶-methyl-guanine methyltransferase encoded by the *ada* gene), (4) excision repair (*uvrA*, *uvrB*, *uvrC*, which make up UvrABC), (5) glycosylase-mediated repair (such as that mediated by AP endonuclease and DNA polymerase I), (6) mismatch repair (in *E. coli*, mediated by *mutS*, *mutL*, and *mutH*), and (7) SOS response (mediated by products of genes: *recA* and *lexA*).

15.

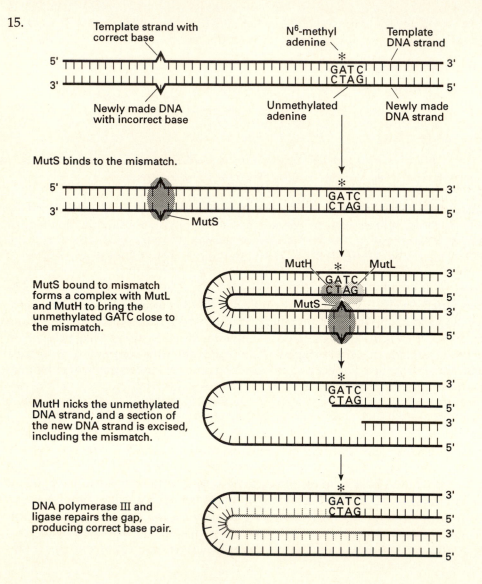

16. A disabled excision repair mechanism for DNA damaged by ultraviolet light is the main defect in people with xeroderma pigmentosum. The dark repair system of *E. coli* is the most similar in that, like UvrABC, the DNA-binding protein defective in xeroderma pigmentosum removes segments of DNA containing distorted base pairs, whether caused by UV-induced dimerization or by covalent modification of bases with large adducts.

17. This hypothesis is well suited to testing by engineering the mutation into the gene by site-directed mutagenesis (SDM), confirming the change has been made by dideoxy sequencing of the gene, and then screening the mutagenized plasmid for its ability to confer resistance to ampicillin. If the amino acid changed is important for β-lactamase (*Amp*^{r} protein) function, you would expect to see a reduced colony size or growth rate in the presence of ampicillin. (Note: the oligonucleotide below is designed to change the histidine to a glutamine at position 5 in the protein. This mutation changes the amino acid chemistry from basic to neutral polar. Other changes that make the amino acid nonpolar, basic, or acidic could be engineered.)

The oligo below would be used to prime *in vitro* DNA synthesis of the plasmid with a DNA polymerase. Ligase would seal the new strand. The replicated plasmid products would then be used to transform *E. coli*. Transformants would be selected using another antibiotic resistance gene (*Tet*^{r}, in this case) before analyzing the plasmid sequence and ability to confer ampicillin resistance.

```
SDM oligo 5' ATG AGT ATT CAA CAA TTC CGT GTC GCC CTT 3'
          5' ... ATG AGT ATT CAA CAT TTC CGT GTC GCC CTT ATT ... 3'
          3' ... TAC TCA TAA GTT GTA AAG GCA CAG CGG GAA TAA ... 5'
             M   S   I   Q   H   F   R   V   A   L   I
```

18. Mutagenize yeast (with a mutagen, such as hydroxyl amine, nitrous acid, or 5-bromouridine, that causes base substitutions), and plate yeast on agar plates with complete medium, including galactose as the sole carbon source. Grow yeast at 25°C until colonies are easily visible. This will select for yeast that are GAL$^+$ at 25°C.

Replica transfer colonies to new plates containing complete medium glucose as the sole carbon source. Grow yeast at 36°C until colonies are easily visible. This will identify the yeast that are *not* temperature sensitive for any essential genes not encoding enzymes necessary for galactose metabolism.

Replica transfer colonies to new plates containing galactose as the sole carbon source. Grow yeast at 36°C until colonies are clearly visible. Identify *all* colonies that grew on the first plate (at the permissive temperature) and the second plate (at the restrictive temperature, but not requiring functional *GAL* genes), but that *didn't* grow on the third plate at the restrictive temperature when functional *GAL* was required. These mutant yeast should be streaked out to purify the strain and then be retested for growth at 25°C and for failure to grow at 36°C when galactose is the sole carbon source.

19. Making a mouse knockout of the β-globin gene requires that we first obtain the mouse β-globin gene. We can do this by screening a mouse genomic library using a probe designed to the human or mouse β-globin gene or, because the cloning of the gene has been published, we could ask the laboratory that cloned the gene if it would send a plasmid containing the mouse gene.

The β-globin gene needs to be internally deleted in a way that will completely destroy its ability to encode a functional β-globin peptide. A common strategy is to replace essential gene material with a selectable marker gene such as *neo*^{r}, a gene encoding resistance to the antibiotic neomycin.

Sometimes the linearized vector (containing the β-globin gene interrupted by *neo*^r) includes a herpes virus gene, *tk*, that encodes thymidine kinase. Thymidine kinase catalyzes a step in the biosynthesis of deoxythymidine, but does it without the selectivity of the host (mouse) version of the enzyme. As a result, thyminelike base analogs can be used to mutagenize (and kill) cells that have taken up the *tk* gene. Why is this useful? If the *neo*-interrupted β-globin gene integrates into the mouse genome by *homologous* recombination, the *tk* gene will be excluded. If the *neo*-interrupted β-globin gene integrates by *nonhomologous* recombination, the *tk* gene will be included. So in this manner, you can enrich your transformants for those that have knocked out a copy of the wild-type gene.

The linearized transformation vector is introduced into mouse embryo cells (or cultured embryonic stem cells) by microinjection or electroporation. The cells are grown on neomycin and in the presence of a base analog that thymidine kinase (if it is also present) will convert to the toxic/mutagenic nucleotide. Embryos that survive are then implanted into a host pseudopregnant mouse. Progeny are screened molecularly (by PCR or Southern blot) to assess the presence of the disrupted gene.

Homozygotes for the knockout can be made by inbreeding the heterozygous mice.

20. There are many. Your answer should include the following:

   1. A trade-off between speed of replication and fidelity of replication is necessary.

   2. Twists and contortions of the double helix and/or tautomeric shifts in the bases that make up DNA lead to occasional mispairing of bases and, subsequently, base substitutions.

   3. Evolution depends on mutation as the ultimate source of genetic variation. A species that eliminated all mutation would eventually be outcompeted by those species that could adapt not only physiologically but genetically.

   4. Sometimes DNA repair is a more urgent need than maintaining sequence fidelity. Whether the DNA damage is due to radiation, nucleases, or transposable elements, broken ends of DNA are very unstable and will be ligated to another reactive DNA end. Even that generates mutations.

23
Extranuclear Genetics

Problems

1. Where within a eukaryotic cell besides the nucleus can genes that can influence the cell's phenotype be found?

2. How does the mitochondrial genome differ from the nuclear genome? Include in your answer comparisons of genome size, structure, gene number and density, types of genes encoded, and replication mechanism.

3. How is the transcription and posttranscriptional processing of genes in the human mitochondrion different from similar nucleus-encoded genes?

4. How would the following codons from an mRNA be translated in the cytoplasm of a human cell? In the mitochondrion? You may assume that the first codon is the site at which translation begins.

    5' AUG AUA GAU UGA AGA 3'

5. Illustrate with four drawings the replication of mitochondrial DNA. Show the initiation of synthesis of the heavy strand, the halfway point for the replication of the heavy strand, the initiation of the light strand synthesis, and the completion of the heavy strand replication.

6. Below are shown sequencing data from the (PCR-amplified) hypervariable regions of mitochondrial DNA from blood and bone samples from living and dead individuals, respectively.

	EXACT BASE POSITION	
	HYPERVARIABLE REGION (HVR) 1	HVR 2
ORIGIN OF SAMPLE	16111 16126 16169 16261 16264 16278 16293 16294 16296 16304 16311 16357	73 146 195 263 315.1
1. Tsarina Alexandra?	T T C C C C A C C T T C	A T T G C
2. Tsar Nicholas?	C C Y C C C A T T T T T	G T T G C
3. child 1/daughter?	T T C C C C A C C T T C	A T T G C
4. child 2/daughter?	T T C C C C A C C T T C	A T ? G C
5. child 3/daughter?	T T C C C C A C C T T C	A T T G C
6. Duke of Fife	C C T C C C A T T T T T	G T T G C
7. Duke of Edinburgh	T T C C C C A C C T T C	A T T G C
8. Anna Anderson	C T C C C C A C C T T T	A T T A -

- = no base at this position, ? = base could not be assigned, Y = C or T.

Samples 1–5 are from five skeletons found buried in a shallow grave containing the remains of six adults and three children near the Russian town of Ekaterinburg. Forensic evidence points to the grave as the possible burial site of the tsar of Russia and his family, physician, and servants, all of whom were murdered early in the Russian Revolution of 1917. Sample 6 is from the duke of Fife, a living relative of the tsar of unbroken maternal descent (see tsar pedigree below). Sample 7 is from the duke of Edinburgh, a maternal grandnephew of Tsarina Alexandra (see tsarina pedigree below). Sample 8 is from Anna Anderson, a recently deceased Russian emigré who claimed to be Anastasia, a long lost daughter of the tsar and tsarina who had miraculously escaped the massacre that killed the rest of her family.

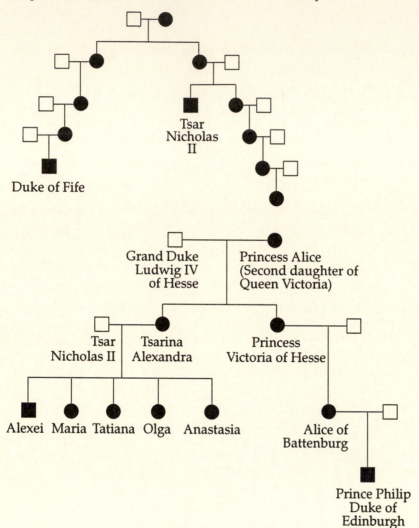

a. Does the mitochondrial DNA evidence from the remains in the grave suggest that the putative tsarina was mother to the children in the grave?

b. Does the mitochondrial evidence show that the tsar was the father of the children in the grave?

c. Is there evidence that the remains in the grave are really those of the tsarina?

    d. Does the DNA evidence suggest that Anna Anderson was Annastasia, daughter of the tsar and tsarina?

    e. What is the significance of the "Y" in the mitochondrial DNA of sample 2, from the skeleton putatively of the tsar?

7. How is RNA editing different from splicing, tailing, and the chemical modification of RNA molecules?

8. What features of mitochondria and chloroplasts suggest that they were both, over a billion years ago, independently living aerobic prokaryotes?

9. Contrast the four distinguishing characteristics of extranuclear inheritance with the characteristics of Mendelian inheritance.

10. Pollen from a pure white branch of a variegated strain of a plant called "marvel of Peru" (*Mirabilis jalapa* strain *albomaculata*) is added to the pistil of a destaminated flower on a variegated shoot. The flower is then covered to protect it from further pollination events. What phenotypes (and in what proportions) are expected in the plants that grow from the seeds from this cross?

11. A branch of a redwood tree (*Sequoia sempervirons*) is albino. Propose a hypothesis to explain this white-needled branch. What experiment could you do to test the hypothesis?

12. A cross is made between mating type "A" protoperithecia of *Neurospora crassa* that are defective for a mitochondrial gene (19S rRNA)—genotype [*mi-1*]—and mating type "a" conidia of *Neurospora* that are wild type for the 19S rRNA gene [*mi-1*$^+$]. The resulting ascospores are dissected out and put in growth medium. What ratio of fast-growing:slow-growing *Neurospora* would be expected from this cross? Why?

13. Describe the phenotypes (*petite* or *grande*) of the progeny (F_1) and mitotic descendants of each of the following crosses:

    a. *pet*$^-$ × *pet*$^+$

    b. [*rho*$^-$*N*] × [*rho*$^+$]

    c. [*rho*$^-$*S*] × [*rho*$^+$]

14. *Chlamydomonas reinhardtii* of genotype *mt*$^+$ *ery*^{s} and *mt*$^-$ *ery*^{r} are mated, and haploid meiotic products are observed. What phenotypes would you expect with respect to erythromycin sensitivity and resistance? Why?

15. The pedigree below shows the inheritance of Leber's hereditary optic neuropathy (LHON) in three generations of a family in which many individuals are affected.

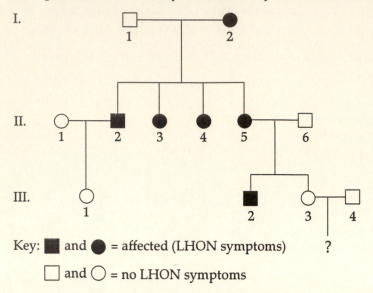

Key: ■ and ● = affected (LHON symptoms)

□ and ○ = no LHON symptoms

a. What is the mode of inheritance of LHON?

b. What is the probability that the first son of the 20-year-old woman III-3 and her normal husband will inherit LHON?

c. Why might the mutation(s) that cause LHON give rise to the death of the optic nerve?

16. URA3⁻ killer yeast of the "a" mating type are combined with TRP1⁻-sensitive yeast of the "α" genotype and plated on uracil-deficient, tryptophan-deficient medium. Of the diploid yeast produced, all were URA3⁺ TRP1⁺, as expected. But all were killer yeast as well. Why?

Challenge Problems

17. In *Drosophila*, there is a mutation, *bcd*, that in the homozygous state has a normal developmental phenotype. Offspring of males that are *bcd/bcd* are also normal. But offspring of *bcd/bcd* females die as embryos, missing both head and thorax. Propose an explanation of the genetics of *bcd*.

18. How might an organism encode for a tissue-specific RNA that differs from the RNA produced by other cells *only* by the presence of a single base difference? In most tissues, the mRNA is UCA; in one tissue, the mRNA contains a UUA stop codon at the same location.

Team Problems

19. Imagine a chromosomal mutation that precisely removes both the Angelman and Prader-Willi genes in the region 15q11–q13. What phenotypes would you expect from the following individuals with this mutation?

a. the son of a man with Angelman syndrome

b. the daughter of a man with Prader-Willi syndrome

c. the daughter of a woman with Angelman syndrome

d. the son of a woman with Prader-Willi syndrome

20. The tabloids periodically contain a story about a person (or tribe) who is photosynthetic. What would some of the genetic difficulties be with the evolution (or creation) of such people?

Solutions

1. Most eukaryotic cells have extranuclear genomes in addition to their nuclear genome. Aerobic animal, plant, and fungus cells have mitochondria that contain circular chromosomes within their nucleoid regions. Plant cells also often have chloroplasts that, like mitochondria, have circular genomes.

   Mutations in mitochondrial or chloroplast genomes are known and can cause phenotypic variation and disease in the organism that has the mutation in the extranuclear DNA.

   There are also organisms that carry bacteria or viruses cytoplasmically in cell types, including cells in the germ line, and then pass these bacterial or viral genes to their progeny. In one type of parasitic wasp, a cytoplasmic virus contributes to the suppression of the immune response of the caterpillar within which the wasp lays its eggs, allowing the wasp larvae to develop inside the caterpillar safe from immunological attack. The developing wasps retain a population of the viruses when they finally leave the caterpillar in which they developed.

2. The mitochondrial genome is much smaller than the nuclear genome. It varies in size among organisms (about 16.9 kb in humans and up to 2,000 kb in some plants). It is circular and typically less dense than genomic DNA. All mitochondria contain about the same amount of unique-sequence DNA and therefore code for very similar RNAs and proteins. Approximately 40 gene products are encoded by the mitochondria, most of them tRNAs and proteins in the oxidative phosphorylation and electron transport pathways. The human nuclear genome contains between 50,000 and 100,000 genes. Mitochondrial DNA replicates not by bidirectional replication forks, but rather by the formation of a D-loop structure and the use of two origins of replication, one for the new heavy strand, one for the new light strand.

3. In the human mitochondrion, transcription is done from two promoters, one from the heavy chain, one from the light. Both transcripts are polygenic, carrying the message for many genes. The two transcripts are cut by enzymes that liberate the tRNAs that are scattered throughout the mitochondrial genome. The tRNAs are modified by adding CCA to their 3' ends. The mRNAs are modified by the addition of poly(A) tails to their 3' ends.

   In the nucleus, most genes are transcribed from gene-specific promoters and are expressed through monogenic transcripts. Many transcripts are spliced and many mRNAs modified by the addition of a 5' cap, neither of which happen in the mitochondrion.

4. cytoplasmic translation: NH_3^+ Met Ile Asp COO^-
   mitochondrial translation: NH_3^+ Met Met Asp Trp COO^-

5.

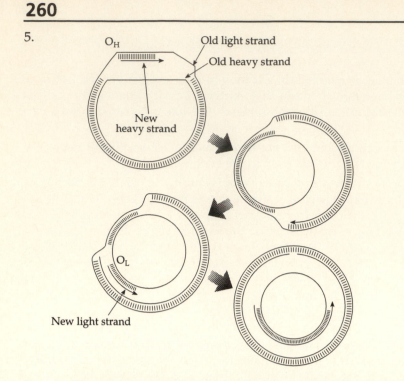

6. a. The mitochondria from the skeleton of the putative tsarina shares bases at all positions in the mitochondrial hypervariable region with each of the children found with her. The evidence supports the hypothesis that she was the mother of the children rather than, say, an unrelated nurse or guardian.

    b. Even if the living relative of the tsar shares the same polymorphic mitochondrial alleles as the children in the grave, there can be no mitochondrial evidence to support the hypothesis that the children are indeed the tsar's children; human males almost never pass on mitochondria to their children. Nuclear DNA typed in experiments not described here showed that the skeleton thought to be that of the tsar shared STR alleles with the nuclear DNA from the skeletons of the children in the grave, establishing paternity.

    c. Because only three skeletons of children were found in the grave (and the tsar had five children), it is possible, but not certain, that Annastasia was among them. Other evidence, such as skeletal morphology, indicates that one son and one daughter are missing, but in the absence of more evidence, it cannot be determined whether Annastasia survived or was buried elsewhere.

    d. The evidence is clear that Anna Anderson was *not* Annastasia. She shares no mitochondrial markers with the living relative of the tsarina (Prince Phillip, Duke of Edinburgh). Because of maternal inheritance, Annastasia would be expected to share most, if not all, of the polymorphisms found in Prince Phillip's mitochondrial hypervariable region.

    e. The Y in the mitochondrial sequencing data from sample 2 suggests that the site contains both T and C in the same individual's mitochondria. That is, the tsar, in this case, had some mitochondrial genomes with a T at base position 16169 and had some mitochondrial genomes with a C at the same position. This is an example of heteroplasmy.

7. RNA editing is the process by which nucleotides are added and/or deleted posttranscriptionally. The editing can be extensive and requires special RNA molecules called "guide RNA" that cleave the transcript, serve as the template for the addition (or deletion) of nucleotides, and ligate the transcript once the deletion or insertion is complete.

   Splicing of RNA is generally catalyzed by snRNAs and associated proteins and refers to the precise removal of intervening RNA from transcripts.

   Tailing of RNA refers to the addition of ribonucleotide triphosphates to the end of a transcript, perhaps after trimming off a 3′ sequence to establish the point of addition.

   Chemical modification neither adds nor removes nucleotides, but rather modifies the bases in systematic ways. A guanine in a tRNA, for example, may have a methyl group added at carbon 1 to make 1-methylguanine.

8. Both mitochondria and chloroplasts have (1) circular chromosomes of their own, (2) a replication mechanism (at least in the case of mitochondria) that is more similar to prokaryotes than eukaryotes, (3) proteins beginning with the amino acid fMet, and (4) ribosomes that are sensitive to the same inhibitors that block prokaryotic protein synthesis.

9. The following are characteristic of extranuclear inheritance:

   1. Reciprocal crosses in which extranuclear genes are examined do *not* give comparable results as they typically do for nuclear genes in somatic cells. Extranuclear genes show uniparental inheritance.

   2. Extranuclear genes cannot be mapped to any nuclear chromosome.

   3. Typical Mendelian ratios (3:1, 9:3:3:1, etc.) are not observed in the F_2 of crosses between true-breeding lines.

   4. Extranuclear inheritance is not affected by the substitution of a nucleus from another cell that has a different extranuclear genotype.

10. Variegation in *Mirabilis* arises from the maternal inheritance of two kinds of chloroplasts. One kind of chloroplast is wild type. The other is called a leukoplast because it is deficient in the ability to make chlorophyll. That is, some chloroplast genomes (cpDNA) have functional genes for chlorophyll-synthesizing enzymes, and others in the same organism do not. Because cells typically have many chloroplasts, those cells that pass on chloroplasts (somatic cells and eggs, but not pollen) will typically distribute some chloroplasts of each type to the daughter cells with each cell division. However, sometimes a cell will get only leukoplasts. This cell will then give rise to a clone of cells in the stem or leaf that are white. Other times a cell will inherit only chloroplasts—with wild-type genes for chlorophyll synthesis. This cell will be the founder of a clone of unvariegated green stem or leaf tissue.

    For this problem, all that is important is the phenotype of the maternal tissue (the eggs in the flower's pistil). Because the pollen contributes no organelles in *Mirabilis,* it is not important for solving this problem. In this case, the pistil was on a variegated shoot. The eggs, therefore, may contain both leukoplasts and chloroplasts, in which case they will become seeds that give rise to variegated plants, with patches of white related to the ratio of leukoplasts to chloroplasts. Other eggs might contain only chloroplasts. These will grow into unvariegated green plants. Eggs containing only leukoplasts will grow only as large as they can with the nutrients stored in the seed and obtainable from the soil. The ratios of each of these outcomes is roughly related to the ratio of chloroplasts to leukoplasts in the shoot bearing the pistil.

11. It is possible that the branch is white because of the loss of a gene in its cpDNA that encodes an enzyme required in chlorophyll biosynthesis. One way to test this hypothesis would be to use pollen from a green branch to fertilize the female cones on the white branch and plant the resulting seeds. If all the plants came up white, one could conclude that maternal mitochondrial inheritance was at work. If the plants came up green or a mixture of green plants and white plants, then it is more likely a nuclear gene that mutated.

    Another way to test the hypothesis that a chloroplast gene is mutated would be to sequence the genes necessary for chlorophyll synthesis in chloroplasts isolated from the white branch tissue and compare it to the same genes in chloroplasts isolated from green branches of the same tree. Mutations that could destroy chloroplast genes necessary for chlorophyll synthesis would be detected in the comparison.

12. [*mi-1*] "female" × [*mi-1$^+$*] "male" will give rise to only [*mi-1*], the slow-growing *Neurospora*. Asci will have 0 fast-growing : 8 slow-growing spores. This is because the parent that forms the protoperithecia is the parent that produces the cytoplasm-rich, mitochondria-containing fruiting bodies that will be fertilized by the small mitochondria-poor conidia.

13. a. *pet$^-$* × *pet$^+$* will yield *grande* heterozygotes of genotype *pet$^-$/pet$^+$*. Mitotic descendants will be *grande*, also. Meiotic products, however, will be half *petite* and half *grande* because nuclear (or segregational) *petites* are inherited in a typical Mendelian manner.

    b. [*rho$^-$N*] × [*rho$^+$*] will yield *grande* colonies that remain *grande* though many mitotic divisions. Meiotic products are also all *grande*. Neutral *petites* are due to the complete loss of the mitochondrial genome in those cells. All the mitochondria are provided by the wild-type strain.

    c. [*rho$^-$S*] × [*rho$^+$*] will yield [*rho$^-$S*]/[*rho$^+$*] yeast that have a respiration capacity intermediate between *petite* and *grande*. Subsequent mitotic division, however, will produce an increasing proportion of yeast that are *petite*. One of two reasons is likely to account for the shift toward *petite* forms. First, it is possible that *suppressive* [*rho$^-$S*] mitochondria replicate faster than normal mitochondria and soon outnumber them in mitotic progeny. Or second, the normal and *suppressive* mitochondria could fuse and, by genetic recombination, severely alter the normal mDNA so that it, too, gives rise to defective mitochondria and *petite* yeast.

14. One would expect most (about 95%) of the meiotic products to be *ery^s* because the *mt$^+$* strain confers the erythromycin sensitivity/resistance character onto the next generation. For an unknown reason, the + mating type generally is the parent that passes on the chloroplasts to the meiotic progeny.

15. a. LHON is inherited maternally, with penetrance proportional to the fraction of the mother's mitochondria that are mutant. Because LHON has a midlife onset, the absence of the disease in a young person such as III-3 (presumably the third generation is young if their grandparents are still alive) is not surprising and not indicative of being unaffected.

    b. The first son of III-3 has a high probability of inheriting LHON, assuming that III-3 has a similar level of heteroplasmy as her mother and mother's mother. Heteroplasmy, however, makes it possible that III-3 has lost the mutant mitochondria or reduced the number of mitochondria carrying the mutation. In either of these cases, the probability that her son will carry the disease would be reduced. DNA testing of the mitochondria of III-3 and her son could give a quicker assessment of both individuals' risks of developing LHON.

    c. The optic nerve, which constantly transmits visual information to the optic centers of the brain using a process dependent on ion gradients in nerve axons that require large quantities of ATP to maintain, has very high energy demands. If these cells are starved for ATP, they will eventually die.

16. The killer yeast phenotype is due to the presence of a cytoplasmic virus of the M type that confers the ability to produce a toxin (and confers resistance to the toxin). When a killer and a sensitive yeast mate, the resulting diploid contains the virus, and so is a killer. All haploid spores will also contain the virus and so will also be the killer type.

17. *bcd* is a maternal effect gene. When defective, it has no effect on the organism carrying it, except that the offspring of females homozygous for the mutation are defective. They are defective because the eggs are not loaded properly with *bcd*+ product, the normal bicoid protein. Bicoid protein is necessary for anterior patterning in *Drosophila*. Without bicoid protein, no anterior (head or thorax) will develop.

18. It is possible to have a tissue-specific expression of a guide RNA (gRNA) that cleaves the RNA, replaces the C with a U, and religates the molecule. As long as the RNA is expressed in only one tissue, that is the only tissue in which the transcript with the change will appear.

19. a. Prader-Willi syndrome 50% of the time, 50% normal

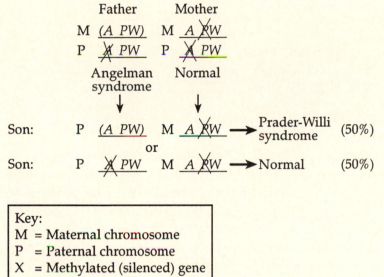

Key:
M = Maternal chromosome
P = Paternal chromosome
X = Methylated (silenced) gene
( ) = Region deleted

b. Prader-Willi syndrome 50% of the time, 50% normal

c. Angelman syndrome 50% of the time, 50% normal

```
              Father        Mother
          M   A P̶W̶     M  (A  PW)
          P   X̶ PW      P   X̶ PW

              Normal      Angelman
                          syndrome
                  │           │
                  ↓           ↓
Daughter:  P   X̶ PW    M (A  PW) ──→ Angelman      (50%)
                                       syndrome
                    or
Daughter:  P   X̶ PW    M  A P̶W̶  ──→ Normal        (50%)
```

d. Angelman syndrome 50% of the time, 50% normal

```
              Father        Mother
          M   A PW      M   A PW
          P   A PW      P   A PW

              Normal     Prader-Willi
                          syndrome
                  │           │
                  ↓           ↓
Son:       P   X̶ PW    M (A  PW) ──→ Angelman      (50%)
                                       syndrome
                    or
Son:       P   X̶ PW    M  A P̶W̶  ──→ Normal        (50%)
```

20. There are, of course, many problems. The following are some of the major ones:

1. If the plant people contain chloroplasts, where did they get them? No simple set of mutations could spontaneously generate in animals all the enzymes necessary for photosynthesis.

2. Even if they could acquire chloroplasts, what about all the nucleus-encoded proteins and RNAs necessary for chloroplast assembly and function? There are many such genes.

3. Photosynthesis biochemistry is optimized for plant cell environments, with the particular concentrations and composition of proteins, ions, gases, carbohydrates, lipids, and cofactors. Animal cells, having diverged evolutionarily from plant cells over a billion years ago, do not have a very similar molecular environment.

4. The expression of plant genes would not occur in animals because the necessary transcription regulatory factors are not the same, or even very similar (except in their basic DNA-binding and protein-protein interaction motifs).

24
Transposable Genetic Elements and Cancer

Problems

1. How are transposition and recombination fundamentally different?

2. Which of the following can result from the insertion of a transposon in bacteria?
   a. gene inactivation
   b. an increase or decrease in transcriptional activity of a gene
   c. deletions or insertions

3. Which of the following can be consequences of transposition in eukaryotes?
   a. conversion of a heterozygote from the dominant to the recessive phenotype
   b. production of a null mutation
   c. increased or decreased efficiency of promoter regions

4. Why does the insertion of an autonomous element result in the production of an unstable (mutable) allele?

5. What is the relationship between a proto-oncogene and its corresponding viral oncogene?

6. Support for the hypothesis that a particular repetitive element is a transposon often comes from analyzing the element's DNA sequence for open reading frames that encode transposaselike products. In humans, *Alu* sequences do not encode any of the enzymes that are presumably needed for transposition.
   a. What aspects of *Alu* structure suggest that it might be capable of retrotransposition?
   b. What is the evidence that *Alu* sequences are capable of moving via an RNA intermediate?

7. Distinguish between a viral oncogene, a cellular oncogene, and a proto-oncogene.

8. Below is an insertion sequence found within a quiescent *E. coli* *nmpC* gene (*nmpC* encodes a protein called porin). Most of the *nmpC* gene has been omitted for simplicity.

```
0681                          gaagccaatg gtgatggttt cggtggtgat gctgccaact
0721 tactgattta gtgtatgatg gtgtttttga ggtgctccag tggcttctgt ttctatcagc
0781 tgtccctcct gttcagctac tgacggggtg gtgcgtaacg gcaaaagcac cgccggacat
0841 cagcgctatc tctgctctca ctgccgtaaa acatggcaac tgcagttcac ttacaccgct
0901 tctcaacccg gtacgcacca gaaaatcatt gatatggcca tgaatggcgt tggatgccgg
0961 gcaactgcgc cattatgggc gttggcctca acacgatttt acgtcactta aaaaactcag
1021 gccgcagtcg gtaacctcgc gcatacagcc gggcagtgac gtcatcgtca gcgcggaaat
1081 ggacgaacag tggggctatg tcggggctaa atccggccag cgctggctgt tttacgcgta
1141 tgacaggctc cggaagacgg ttgttgcgca cgtattcggt gaacgcacta tggcgacgct
1201 ggggcgtctt atgagcctgc tgtcaccctt tgacgtggtg atatggatga cggatggctg
1261 gccgctgtat gaatcccgcc tgaagggaaa gctgcacgta atcagaaagc gatatacgca
1321 gcgaattgag cggcataacc tgaatctgag gcagcacctg gcacggctgg gacggaagtc
1381 gctgtcgttc tcaaaatcgg tggagctgca tgacaaagtc atcgggcatt atctgaacat
1441 aaaacactat caataagttg gagtcattac cggtttcggt ttctcccacta cttatgagta
```

a. Find and label the target site duplications.

b. Find and label the two terminal repeats (inverted with respect to each other) of the insertion sequence.

c. How long is the insertion element? Exclude the duplicated target sequence.

d. What family of insertion sequences is this IS in?

e. Draw the 15 bases extending in both directions from the edge of the target site, and label where the DNA strands were cut when the insertion sequence integrated into the genome.

9. a. What is an acute transforming retrovirus?

b. What are the main requirements of its life cycle?

c. What is the role of the virus in tumorigenesis?

10. Studies of retinoblastoma have contributed tremendously to our understanding of tumorigenesis.

a. What is retinoblastoma?

b. How is it transmitted?

c. What kind of protein does the *Rb* gene encode?

d. What are some of the mechanisms by which the gene can be lost?

e. Describe the model of tumorigenesis that arose from a study of retinoblastoma.

11. What is the difference between a tumor cell and a cancer cell?

12. What are the three major classes of genes that, when mutated, contribute to the progression of tumorigenesis? What additional classes of genes are mutated in cells that have become cancerous?

13. On the schematic diagram of the human immunodeficiency virus (HIV) below, label the following:
    a. the viral genome
    b. the product of the *POL* gene
    c. the product of the *GAG* gene
    d. the product of the *ENV* gene
    e. the membrane from the infected cell

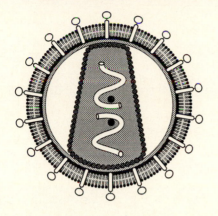

14. How is a transforming retrovirus different from a normal retrovirus?

15. What is the normal role of the product of the tumor suppressor gene, *p53*?

16. What is the difference between a retrovirus and a retrotransposon? Give an example of each.

Challenge Problems

17. Barbara McClintock worked for years studying unusual inheritance patterns in maize. She discovered *transposable elements* during her study of unstable mutant alleles (mutations that revert to wild type unusually frequently) in corn kernels. Each corn kernel has an embryo that is diploid (originates from one sperm and one egg) and a triploid endosperm that originates from one sperm (n) and two polar nuclei ($n + n$). The color of the kernel is determined by the endosperm genotype.

    The *C* gene encodes an enzyme that produces a pigment. *c* is a base-pair substitution mutation that makes a white kernel when homozygous. c^{Ds} is an unstable allele caused by the insertion of the Dissociation element (*Ds*). Activator (*Ac*) is absent from Ac^+ chromosomes. Assume the Activator and *C* loci are unlinked. Transposition can happen anytime during the embryo's or kernel's development.

    What kernel phenotypes (solid color, colorless, spotted) would be produced, and in what ratio, when the following cross is performed?

    (pollinating plant) $C\ c^{Ds}\ Ac\ Ac^+ \times c\ c\ Ac^+\ Ac^+$ (plant supplying egg and polar nuclei)

18. Draw a schematic illustration showing how two integrated copies of bacteriophage Mu can lead to deletions in the *E. coli* genome between them.

Team Problems

19. Why are cancer researchers so interested in finding people with cancer who also have any of the following, even if most people with the particular cancer *don't* have these situations:
    a. a large family with many people who also have the same cancer
    b. a known chromosomal aberration
    c. syndrome of other problems

20. Retroviruses have many compelling qualities as vectors for gene therapies. There are also some significant concerns regarding their use.
    a. What are some of the qualities that make retroviruses well suited for gene delivery?
    b. What might be the basis of some of the concerns?
    c. How could retrovirus-inspired gene therapy vectors be made better?

Solutions

1. Most recombination events depend on (and require) DNA homology; most transposition does not.

2. All of these can be caused by transposition in bacteria.

3. All of these can be caused by transposition in eukaryotes.

4. An autonomous element generally mutates the gene it integrates into by interrupting an exon, causing a frameshift (and premature stop codon) or separating sequences that need to be near each other. Autonomous elements can excise, however, precisely to yield a complete revertant and imprecisely, thereby causing a stable deletion or insertion.

5. The proto-oncogene encodes a product essential to the normal development and function of the organism. The viral oncogene encodes an altered product that has aberrant function.

6. a. *Alu* repeats have an internal promoter. *Alu* sequences appear to have evolved from processed copies of 7SL RNA, transcribed by DNA polymerase III from their internal promoter (i.e., the transcript takes the promoter with it; it is *not* upstream of the start of transcription).
    b. Evidence for *Alu* mobility includes the presence of direct repeats at ends of *Alu*, and experiments that have shown that *Alu* can still be transcribed by DNA pol III, generating RNA copies of *Alu*. If reverse transcriptase is available in the cell, it is not difficult to imagine at a new site the generation of *Alu* DNA fragments that reintegrate into the genome, adding to the abundance of *Alu* sequences.

7. A viral oncogene is a partial or complete copy of a proto-oncogene from the host that, because of its truncation (and, for example, loss of regulatory sites) or because of its high, unregulated level of expression, is capable of driving cell growth and division in cells infected by the transforming virus that contains the viral oncogene.

A cellular oncogene is a partial or fused copy of a proto-oncogene that, due to mutation, has lost its normal regulated expression or regulatable function. A cellular oncogene contributes to unregulated cell growth and division. It is *not* part of a transducing virus.

A proto-oncogene is a gene that can be mutated into a gene that drives unregulated growth and cell division. This normal, wild-type version of what is usually considered a "cancer gene" is essential for normal, regulated progression through the cell cycle in nontumor cells. It is also *not* part of a transducing virus.

8. a. The target site duplications are underlined in the sequence below.

    b. The inverted terminal repeats are italicized in the sequence below. Note: the inverted terminal repeats are not perfectly complementary.

```
0681                             gaagccaatg gtgatggttt cggtggtgat gctgccaact
0721 tactgattta gtgtatgatg gtgtttttga ggtgctccag tggcttctgt ttctatcagc
0781 tgtccctcct gttcagctac tgacggggtg gtgcgtaacg gcaaaagcac cgccggacat
0841 cagcgctatc tctgctctca ctgccgtaaa acatggcaac tgcagttcac ttacaccgct
0901 tctcaacccg gtacgcacca gaaaatcatt gatatggcca tgaatggcgt tggatgccgg
0961 gcaactgcgc cattatgggc gttggcctca acacgatttt acgtcactta aaaaactcag
1021 gccgcagtcg gtaacctcgc gcatacagcc gggcagtgac gtcatcgtca gcgcggaaat
1081 ggacgaacag tggggctatg tcgggctaa atccggccag cgctggctgt tttacgcgta
1141 tgacaggctc cggaagacgg ttgttgcgca cgtattcggt gaacgcacta tggcgacgct
1201 ggggcgtctt atgagcctgc tgtcacccctt tgacgtggtg atatggatga cggatggct
1261 gccgctgtat gaatcccgcc tgaagggaaa gctgcacgta atcagaaagc gatatacgca
1321 gcgaattgag cggcataacc tgaatctgag gcagcacctg gcacggctgg gacggaagtc
1381 gctgtcgttc tcaaaatcgg tggagctgca tgacaaagtc atcgggcatt atctgaacat
1441 aaaacactat caataagttg gagtcattac cggtttcggt ttctccacta cttatgagta
```

    c. The IS element here is 767 bases long (1472 – 705).

    d. This insertion sequence is in the IS1 family.

    e. The target site in the *nmpC* gene, and where the site was cut, are shown below.

```
                            cut
                             |
      5' GAAGCCAATGGTGAT GGTTTCGGT TTCTCCACTACTTAT 3'
      3' CTTCGGTTACCACTA CCAAAGCCA AAGAGGTGATGAATA 5'
                           |
                          cut
```

9. a. Acute transforming retroviruses (ATR) are retrovirus particles that transform a host cell rapidly and with high efficiency.

    b. For the ATR to replicate, it must infect a cell that has also been infected by an intact retrovirus that then provides the proteins not encoded by the ATR. The virus gains access to the cell (by endocytosis), reverse-transcribes their RNA genome, copies the DNA strand to become dsDNA, integrates into the host DNA, transcribes viral genes, makes viral proteins, and assembles virus particles, which then exit the cell.

    c. The ATR genome includes an additional gene—an oncogene—that usually replaces one or more essential viral genes, making the virus' replication defective. The viral oncogene causes the loss of growth control, which gives rise to the tumor (mass of proliferating cells) and contributes to the accumulation of mutations because the coupling of DNA repair and entry into S phase may be disrupted. The normal helper virus benefits from the oncogene-induced host cell proliferation.

10. a. Retinoblastoma is a malignant tumor composed of primitive retinal cells usually occurring in children less than 3 years old. Retinoblastoma occurs in hereditary forms. In the case of hereditary retinoblastoma, affected individuals generally have tumors on both eyes.

   b. Retinoblastoma is transmitted as a dominant trait. That is, inheriting one defective copy *predisposes* one to having the tumors. But both copies of the gene must be destroyed in a cell for the tumor to form.

   c. The gene involved (*Rb*) is a tumor suppressor. pRb, a nuclear protein, inhibits transcription factors E2F, which govern the progression into S phase. When pRb is phosphorylated (by cyclins, etc.), it releases the E2Fl.

   d. The mutation can be lost via point mutation, deletion, chromosome loss, uniparental disomy, mitotic recombination, mitotic nondisjunction, and in other ways.

   e. The two-hit hypothesis arose from work on retinoblastoma. The idea is that inheriting the first hit (a single mutation in *Rb*) sets one up for tumor formation. Later in life (usually in the first 3 years), a second mutation occurs, resulting in a cell that totally lacks the pRb protein. This cell is then the founder of the tumor. Given that both eyes are entirely composed of cells of the *Rb*/+ genotype, all those cells are candidates for a second *Rb* mutation, and hence tumor formation. The two-hit hypothesis likely applies to many diseases where there is an underlying genetic predisposition rather than a completely penetrant, identically expressive phenotype.

11. A tumor cell has lost growth control and fidelity of DNA replication. A tumor is benign to the extent that it is not life-threatening—usually because it grows slowly and stays at its point of origin. A cancer cell is a malignant tumor—a cell mass that threatens the life of the individual carrying it. In addition to being uncontrolled, it is characterized by the acquisition of another feature: metastasis—the spreading of tumor cells to tissues distant from the tissue of origin.

12. Major genes involved in tumorigenesis include (1) tumor suppressor genes, (2) oncogenes, and (3) mutator genes. Mutations in these release the cell from stringent growth control mechanisms, induce the cell to grow and divide, and increase the mutation rate in the tumor cell.

   Cancer cells have mutations in each of the classes of genes described. They also have *some* of the following kinds of metastasis-promoting mutations: mutations that allow the cancer cell to be released by its tissue of origin, enter the bloodstream, exit the bloodstream, avoid or resist the cells of the immune system (that would destroy these cells found out of their proper context), and recruit blood vessels to the invasion site.

13.

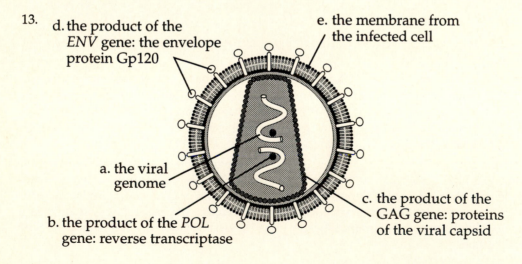

d. the product of the *ENV* gene: the envelope protein Gp120

e. the membrane from the infected cell

a. the viral genome

b. the product of the *POL* gene: reverse transcriptase

c. the product of the *GAG* gene: proteins of the viral capsid

14. The transforming retrovirus includes a gene, obtained at some point from the host, that promotes mitogenesis, or cell division, when the gene is expressed. When a transforming retrovirus integrates into the host genome, it promotes cell division via the viral oncogene it contains. Frequently, the transforming virus is defective for an essential viral gene, however. In such cases, other wild-type viruses are needed as *helper viruses* that encode the necessary packaging proteins and genome replication enyzmes.

15. p53 is a DNA-binding protein and transcription factor. Cell cycle regulators interact with p53 to halt progression through the cell cycle past G_1 when conditions are not right, such as when the cell has just suffered DNA damage. If the DNA damage is great, p53 is able to switch the cell to the suicide pathway, a process called programmed cell death. p53 is centrally important in protecting us against cancer and is frequently inhibited by DNA tumor viruses and deleted in cancer cells.

16. A retrovirus is an RNA virus that, upon entering its host cell, generates a double-stranded copy of its genome using its reverse transcriptase and cellular enzymes such as DNA polymerase and ligase. The retrovirus integrates into the genome and can lie latent there for years. When it excises, however, it is able to escape the cell through the membrane by encoding proteins that comprise a capsid (to contain the RNA genome and reverse transcriptase molecules) and envelope proteins (sent to the cell membrane to await budding of the capsid). HIV is a retrovirus.

    Retrotransposons include retroviruses, but additionally include any transposon that works through an RNA intermediate even if it lacks genes for proteins to encapsulate it and allow it to leave the cell it resides in. Sequences in the *Alu* family of mobile elements are retrotransposons.

17.

X	Polar bodies		
Pollen	c c Ac^+ Ac^+		
C Ac	C c c Ac Ac^+ Ac^+	$\longrightarrow$ Solid color	
C Ac^+	C c c Ac^+ Ac^+ Ac^+	$\longrightarrow$ Solid color	$\Big\}$ 1/2
$C^{Ds}Ac$	c^{Ds} c c Ac Ac^+ Ac^+	$\longrightarrow$ Spotted	1/4
$C^{Ds}Ac^+$	c^{Ds} c c Ac^+ Ac^+ Ac^+	$\longrightarrow$ Colorless	1/4

18.

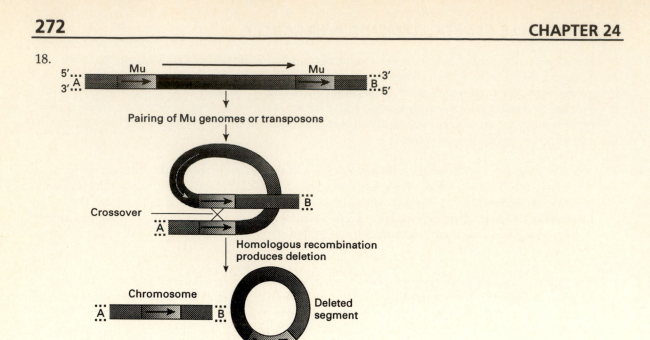

Deleted segment and chromosome
each contain one Mu genome or transposon

19. Though it might seem most useful to study people who have the most standard form of a disease, there are many advantages for a geneticist in finding people who fit any of the three descriptions given in the problem.

   a. A large family with many people affected makes possible the tracking of the gene by identifying regions of an ancestor's genome that are almost always associated with individuals who have the cancer and not found in those in the family who don't have it. Narrowing down the location of the gene is quicker when there are many affected relatives.

   b. Chromosomal aberrations found in someone who has a particular disease, such as cancer, are interesting because they *might* be the cause of the disease for the individual who has it. Although most people might have the disease because they have two point mutations (invisible in a karyogram), one person who has the cancer because of a small deletion or inversion can advance the search for the gene dramatically because one simply needs to look at the break points of the inversion or translocation, or the deleted region, to find candidate genes.

   c. Syndromes are interesting to geneticists because they often suggest that several genes are deleted (or duplicated) in a small chromosomal aberration. As in part b, finding the small deletion or duplication allows the researcher to focus her search on a very small part of the genome, speeding the search for the gene that, when damaged, predisposes one to the cancer.

20. a. 1. Retroviruses integrate into the genome of the host, providing a perhaps lifelong delivery of the therapeutic agent.

       2. Retroviruses do not induce as severe an immune response as other gene therapy vectors, such as adenoviruses.

       3. Once retroviruses are in the body, they deliver DNA to cells efficiently.

       4. Retroviruses preferentially infect dividing cells, so they would be particularly well suited to target cancer cells.

   b. 1. The very high numbers of retroviruses that would be necessary for gene therapy applications cannot easily be made in the laboratory.

       2. Retroviruses infect only actively dividing cells, making them unsuitable for delivering gene therapies to cells such as neurons, which don't divide.

       3. Safety issues are, of course, very important to resolve. For example, could the gene therapy vector recombine with another retrovirus or endogenous gene in the patient and lead to an infection with a potentially dangerous newly constituted virus? Might the integration of a retrovirus vector mutate the host genome, leading to other problems subsequent to the gene therapy?

       4. Patient willingness to be given a virus, even a highly engineered one, is also an important concern.

   c. 1. Ways of increasing retrovirus titer could be studied.

       2. The vector could be designed to integrate with high specificity to a site away from other genes.

       3. The vector could be stripped down to the absolute minimal "virus content," which would make it less likely that the gene therapy vector would cause other problems.

Given the wealth of genetic information currently being discovered, learning how genetic information is encoded and decoded, transferred, stored, and modified is not only tremendously exciting, but is also likely to be the foundation of some of the most powerful new technologies of the twenty-first century.